'*Women in Vinyl* is an essential resource for anyone desiring to work in the vinyl industry, pushing outside the stereotypical record collector to showcase role models working in the field. It also provides education around physical media, something rarely taught in traditional school settings. Anyone wishing to learn more about the inner workings of this industry will find this to be their "go-to" starting point!'

Rain Phoenix, *Performing Artist/LaunchLeft founder*

'If you are thinking about a career in the vinyl record industry, or want to learn more about the processes that go into manufacturing those records that we love, then this book is a great resource for you. *Women in Vinyl* showcases ground-breaking women and explores how they got into this exciting field. These educational chapters and collections of interviews will be an invaluable read for those searching for insight into this field.'

Jim Eno, *Spoon*

'*Women in Vinyl* is a great resource to learn about different careers in vinyl from a variety of professionals working in the field. It's clear that this community loves what they do, wants to share what they know, and support others getting into it.'

April Tucker, *author,* Finding Your Career in the Modern Audio Industry

'You'd think that I would have a pretty good understanding of how that PVC groove makes its way to the turntable. After all, I've had pretty close to a 24/7 relationship with it most of my life: I hotly debated matrix messages; I cut, glued and stuffed 7" Minor Threat sleeves in my basement; I lugged countless 30 count boxes of 12 inchers. Turns out I knew very little. Thankfully, this great read not only explains everything about vinyl, it is a righteous and joyous celebration of the dedicated women who have been and will keep on mastering, cutting, pressing and delivering those sublime sonic platters!'

Lyle Preslar, *Minor Threat*

'*Women in Vinyl* does an important service to the music industry by naming and spotlighting the stories of female professionals in the vinyl manufacturing sector. The book creates awareness as well as showing that everyone's journey in the music business is unique. A great resource and a fun read!'

Portia Sabin, *President, Music Business Association*

Women in Vinyl

Women in Vinyl: The Art of Making Vinyl provides a comprehensive guide to the world of vinyl, with a focus on empowerment, diversity, and inclusion, designed to both demystify the vinyl community and highlight the vital role women and minority groups play in shaping the industry.

Divided into each step of the process, the book provides a detailed overview of the vinyl manufacturing process, from lacquer cutting, electroplating, and record pressing, to the roles of record labels, distribution, DJs, and more. With interviews and profiles from global professionals throughout, the book is a first-of-its-kind guide to the vinyl industry and the women who are blazing trails within it.

Women in Vinyl is an essential resource for professionals, hobbyists, and students interested in the process of making vinyl, including those who want to deepen their understanding of the vinyl medium and its role in shaping the music industry, as well as for those interested in the work of the organization Women in Vinyl.

Jenn D'Eugenio is the Founder and President of the non-profit organization Women in Vinyl. As an avid record collector for over two decades, her passion for vinyl records has led her to a career working at record pressing plants to help manufacture the physical product for bands, and major to independent record labels.

Women in Vinyl

The Art of Making Vinyl

Jenn D'Eugenio

LONDON AND NEW YORK

Designed cover image: Miloje/Shutterstock.com

First published 2024
by Routledge
4 Park Square, Milton Park, Abingdon, Oxon OX14 4RN

and by Routledge
605 Third Avenue, New York, NY 10158

Routledge is an imprint of the Taylor & Francis Group, an informa business

© 2024 Jenn D'Eugenio

The right of Jenn D'Eugenio to be identified as author of this work has been asserted in accordance with sections 77 and 78 of the Copyright, Designs and Patents Act 1988.

All rights reserved. No part of this book may be reprinted or reproduced or utilised in any form or by any electronic, mechanical, or other means, now known or hereafter invented, including photocopying and recording, or in any information storage or retrieval system, without permission in writing from the publishers.

Trademark notice: Product or corporate names may be trademarks or registered trademarks, and are used only for identification and explanation without intent to infringe.

British Library Cataloguing-in-Publication Data
A catalogue record for this book is available from the British Library

Library of Congress Cataloging-in-Publication Data
Names: D'Eugenio, Jenn, author.
Title: Women in vinyl: the art of making vinyl / Jenn D'Eugenio.
Description: Abingdon, Oxon; New York: Routledge, 2023. | Includes bibliographical references and index.
Subjects: LCSH: Women in the music trade. | Sound recording industry. | Sound recordings—Production and direction.
Classification: LCC ML3790 .D45 2023 (print) | LCC ML3790 (ebook) | DDC 781.49082—dc23/eng/20231108
LC record available at https://lccn.loc.gov/2023027452
LC ebook record available at https://lccn.loc.gov/2023027453

ISBN: 978-1-032-35093-6 (hbk)
ISBN: 978-1-032-35092-9 (pbk)
ISBN: 978-1-003-32537-6 (ebk)

DOI: 10.4324/9781003325376

Typeset in Rotation
by codeMantra

Contents

Acknowledgments ix
About the Author x
Foreword xii
Carrie Colliton

Introduction: About Women in Vinyl 3

1 **Lacquer Cutting** 7
  Jett Galindo
  Amy Dragon
  Heba Kadry
  Margaret Luthar
  Mandy Parnell

2 **Electroplating** 31
  Yoli Mara
  Janine Lettmann
  Desiree Oddi
  Elsie Chadwick
  Emily Skipper

3 **Manufacturing** 57
  Caren Kelleher
  Anouk Rijnders
  Ren Harcar
  Brianna Orozco
  Karen Emanuel

4 **Distribution** 83
  Amanda Schutzman
  Jocelynn Pryor

Christie Coyle
Shelly Westerhausen Worcel

5 Record Labels — 103
Désirée Hanssen
Katy Clove
Riley Manion
Katrina Frye
Julia Wilson

6 Record Stores — 127
Lolo Reskin
Brittany Benton
Sharon Seet
Claudia Wilson
Shirani Rea

7 Lathe Cutting — 151
Robyn Raymond
Bailey Moses
Tasha Trigger
Emily Nobumoto
Oihane Follones

8 DJing — 175
Colleen "Cosmo" Murphy
Misty Fujii
Monalisa Murray
Dana Brown
DJ Honey

Glossary — 199
Photo Credits — 203
References and Further Reading — 207
Index — 209

Acknowledgments

Enormous thanks to my dear friend, Amanda McCabe, for editing this book and helping me sound better than I could have on my own. To my husband Ray Blevins for being a constant supporter, champion, and ally. My mom and sister Sara, for being unwavering cheerleaders in any pursuit I take on. To my dad, whom I miss dearly; he always supported my drive for anything I put my mind to and would have thought this was all "so neat." Thank you to Robyn Raymond and Jett Galindo for their hours, ideas, inspiration, leadership, and passion for our mission and for being such trailblazers in their respective careers.

Thank you to everyone who was interviewed for and has participated in this book, you are making a huge difference, and we hope this will shine a little more light on your hard work.

Thank you to Carrie Colliton for being a leader, writing our foreword, believing in us, and championing our platform. Sincere appreciation to all our friends who fact-checked their respective chapters, making sure we were as honest and thorough as possible: Jett of the Bakery in LA; Robyn of Red Spade Records; Carrie of Record Store Day, and the Coalition of Record Stores; Ray, DJ Rayblev and Production Director at Gold Rush Vinyl; Amanda Schutzman of All Media Supply; Martin Frings of Stamper Discs; Heath Gmucs of Wax Mage Records, and Gotta Groove for his innovation. As well as Sarah and Levi Seitz, Owner and talent of Black Belt Mastering; Yoli and Chris Mara, Owners of Welcome to 1979; the Lathe Lord Mike Dixon, one of the coolest Scott Lamb of Pheenix Alpha; long time WiV friend Scott Orr of Other Record Labels; and last but never least the guy who does it all while looking stylish and sipping on the best wine and cocktails, Jason Bitner of Traffic Entertainment.

To our board and advisors for lending us their passion and dedication to the medium and our goals. There are numerous people involved in the process of creating and how we enjoy our records. So thank you to all of you who bring us a part of the experience that we couldn't cover here; the printers like Rob and the crew at Stoughton, the archivists preserving our history, and storage companies like Kate Koeppel. Also, thank you to Ali and Eric at Furnace Record Pressing for giving me the job, and seeing the passion and potential I had for it.

To all those working hard every day at record stores, at pressing plants, those cutting or plating, the people cheering us on from the sidelines, the collectors, clients, customers, sponsors, collaborators, and all that believe in our cause and help keep this going, thank you! Truly.

Last but certainly not least, all the amazing humans we have interviewed for Women in Vinyl from our very first blog post until now. We wish we could have included all of you in this book (maybe in Part 2); without each of you, none of what we are working to do would be possible. You are role models, not only to us and to each other but to the next generation of the vinyl industry.

About the Author

As a child, when I wanted a book in the library, it would have been about dinosaurs, a tomboy ballerina's journey, or a "choose-your-own-adventure" book. Writing this now to celebrate women's accomplishments and create role models feels like a full-circle moment. I've been drawn to various forms of art throughout my life, starting ballet at age 3; I fell in love with it and continued to practice for the next 18 years. From then on, my life has always had a soundtrack. I hung out in high school listening to friends' bands and driving around in my car simply to listen to my favorite CDs. I always found ways to incorporate music into school assignments, like designing fake album covers, writing papers dissecting my favorite rock songs, or creating a board game around Woodstock (true story). Yet, somehow I never considered there was a place for me in the music industry.

After graduating from art college with a degree in textile design, where I learned the essential skills of weaving on a loom, dying fabric with natural materials, spinning yarn, and making repeat patterns, I found a job in the design industry. It may sound like I'm joking about these being important skills, and I do poke at it humorously, but these skills taught me about passed-down traditions, cultures outside of my own, and the importance of art and design's impact on our everyday lives.

I took those skills and worked for Fortune 500 companies designing children's clothes. It was some of the most fun I have had in a job, creating happy, bright imagery from a child's perspective. Being creative day-in-and-day-out eventually took a toll, and I found myself looking for a career change.

It's never too late to change directions in life, particularly when fulfillment becomes elusive. I wound up in higher education as a career adviser for my alma mater. There, I helped students in the school of design start their careers. When I say the previous job was the most fun, this was the most fulfilling. I've never experienced anything that felt more rewarding. That said, the lingering love for music was still strong and apparent as I lugged my records across the country, moving back and forth. My mom knew it was something I loved, and in 2017 she told me that a vinyl record-pressing plant was opening in my hometown. My husband is also a collector and in the industry; it seemed like the perfect opportunity. We packed our bags and started this next chapter together.

That last career change allowed me to discover the perfect job I never knew I wanted. Today, I've found a way to combine all these different life pursuits into one constant love, music. Working in record pressing, I help artists, bands, and labels make their creative ideas a reality.

With Women in Vinyl, I help empower women and underrepresented people in this industry to achieve their goals and find a career path here on purpose, not by chance. I get to utilize my design experience, love of advising, and passion for passed-down history in my favorite archival format, vinyl.

I hope that everyone picking up this book feels inspired by these stories. And that these stories help you find your fire to blaze a trail in this industry and combine your experiences to create the thing you are most passionate about.

Foreword

I moderated a panel called Women in Vinyl a few years ago at a vinyl industry conference. The panelists were owners/executives from labels, pressing plants, recording facilities, and packaging companies, all women, as the panel title implies. On one of the pre-panel calls, as we got acquainted and worked through talking points, we semi-joked about walking up on stage, where I would then say: "We're women. We work in vinyl." Mic. Drop.

To be clear, this wasn't the only panel with female panelists, and I don't think the male organizers were consciously trying to make any "statements." Still, I came out of it with a slight sense that we were being pulled aside from colleagues and singled out for one aspect of our being. As if we were lumped together and mandated with telling the tale of women in a manufacturing-based, historically assumed-to-be-male industry. Turns out we ALL HAD DIFFERENT STORIES! (Shocking, I know.) Some of us have experienced things or comments so cliché that I'm rolling my eyes just thinking about them. A few dealt with obstacles a male would not have faced. At least one, though, had not experienced any of that ("I literally don't ever think that's been an issue" – I paraphrase, but … BADASS!). We all had unique paths to a career in this revitalized industry and important roles to play in its revitalization. Did we NEED a "Women in Vinyl" panel to brand us?

I was a girl in vinyl. I got my own turntable one birthday when we had just finished moving again. (Military brats unite!) You know how parents always pick something up 'for the kids' when they travel for work? My father decided that the best presents for me were records he got wherever he was – my little stash grew. My brothers and I can still sing songs from the records my mother enjoyed. Eventually, the turntable from the fancier family stereo came with me when I went away to school. I'm happy to admit components of that stacked system still figure in the setup I'm listening to right now. In college, I worked in record stores (I filled our living room with a project to create fuzzy fabric letters taller than I am to spell out VINYL on the back wall of one). I also worked in radio (both college and commercial, flipping and cueing wax at both). I've always had a vinyl collection as part of my adult interior design aesthetic. Today I live in a house with another vinyl collector, one television but three working turntables, and woefully inadequate shelf space. I've never stopped working for record stores, and a whole lot of them never stopped selling vinyl. I helped create a holiday for them, Record Store Day, that markedly helped bring the format back to life. I work in glorious Haim/Highwomen harmony with women at all industry levels. I've also had infuriating conversations and work experiences because I'm a woman.

I have, inarguably, a very cool job in a very cool industry. Some aspects of my career have been made more difficult by, or greatly benefitted from, me being the gender I am. It's all wrapped up together, and it's my one-of-a-kind story. And here's where we come to the twist in my way of thinking about the phrase "Women in Vinyl." The beauty of this book (and the website, podcast, and social media that share its name and creators) is that it doesn't "lump"; it illuminates. It doesn't "single out"; it spotlights. It showcases people who are integral to one of the most fascinating industries in business and pop culture. I imagine you'll read through it and be so interested in their stories and what goes into their jobs – SO MANY VERY COOL JOBS! You won't be thinking about that W word in the title that serves as the connective tissue between them. And when it does strike you that all these people designing, mastering, cutting, pressing, packing, shipping, flipping, slinging, and playing the vinyl you love are women? Re-open the book and meet them all again.

We're women. We work in vinyl.

Mic. Drop.

Carrie Colliton

Co-Founder, Record Store Day

Director of Marketing/Dept of Record Stores

INTRODUCTION

ABOUT WOMEN IN VINYL

Our mission: to empower women, female-identifying, non-binary, LGBTQ+, BIPOC, and otherwise marginalized humans working in the industry to create, preserve, and improve the art of music on vinyl.

In 2018, a meme was going around social media of a 1950s couple, the man holding a record and the woman knitting. The man's speech bubble shows him sharing the intricate details of the particular version of the album he's holding, to which his wife responds that she could care less. It didn't take long before I noticed an endless number of men sharing this meme to represent their relationship experiences. I became increasingly frustrated by the antiquated boys club represented by this simple image. It was a seemingly innocent laugh that perpetuated and spoke volumes about an out-of-date and harmful bias cutting into the vinyl-collecting community. If you were listening to conversations in my home, you would quickly see that I resonate with the man in the picture. I love learning about pressing variants; luckily, my partner and I enjoy sharing these conversations. I saw little to represent my experience or my place in the community. A lingering feeling of frustration stuck with me.

I started working at Furnace Record Pressing in sales and customer service that same year. Working at a pressing plant was the dream job I never knew I wanted; it was the perfect opportunity to combine my passion for design, music, and helping people. As I met more people in the vinyl industry, I saw how many women held leadership positions across every aspect of the business. Women were opening and running successful pressing plants and cutting **lacquers**. It's not just manufacturing; every step of the vinyl record supply chain benefits from women leading and innovating both at the helm and behind the scenes.

Having been a record collector myself for over 20 years, I had occasionally felt the impact of gender bias firsthand when my husband and I went into record stores together. Sometimes I was treated as if I didn't belong or didn't know and respect the records the same way my husband or the male customers did. Then I noticed this bias creep into conversations with some of our clients as they questioned my understanding of a specific process or downplayed the value of a suggestion I made.

These concurrent events inspired me to create Women in Vinyl as a community resource to disrupt what I saw in the industry and bring women's voices to the forefront. My experience working in higher education for an art college heavily influenced me to envision the organization as an educational non-profit. I love mentorship and celebrating the success of others; I know how powerful it is to support other people's work and how rewarding it is to work together to achieve a goal. I wanted to create a way to give back, be an educational resource, and help people find their way to this industry on purpose and not by chance.

I have always loved music; I remember finding my parents' records and looking at the large-scale artwork and being enamored. I would visit my uncle and drag him to every possible CD store before I had to leave, but there was never quite enough time to hit them all. And I spent hours of my life in high school hanging out watching my friends' jam sessions. However, not being a musician, it never occurred to me that there was a place for me in the industry. College prep offices didn't showcase these music industry opportunities in their pamphlets. Let's be honest, maybe I'd have tried harder in math class if I'd realized it could translate into me cutting records for my favorite musicians. It's time for this to change.

Women in Vinyl has grown from its humble beginnings as a blog and online community focused on showcasing and highlighting women's accomplishments to be so much more. We achieved non-profit status in 2021 and continue to find new ways to expand how we inspire and create new role models. We support all minority populations in the vinyl industry, from cutting engineers, record labels, pressing plants, DJs, and those running or working in record stores. Women in Vinyl showcases the individual successes of community members and provides a non-judgmental platform for building connections and professional networks. We are a nexus for sharing their experiences and advice to propel the industry forward. We support career development, and provide tools to help those working in the field advance in their career, expand opportunities in these fields, and share resources and knowledge through demystification, education, collaboration, and diversity. Thank you for picking up this book and participating in this change.

CHAPTER ONE

LACQUER CUTTING

Lacquer, as defined by Merriam-Webster (n.d.), is "1. (n) any of various clear or colored synthetic organic coatings that typically dry to form a film by evaporation of the solvent or 2. (v) to coat with or as if with lacquer." When it comes to getting records into your hands, let's think of "a lacquer" as a unique combination of these definitions.

Starting their life as relatively unremarkable-looking aluminum discs, "lacquers" are the tangible element between recorded sound and what will become your record. They come in standardized sizes, just a little bigger than the record size they help create. For example, a 14" lacquer will become a 12" album, a 12" lacquer is used to make your 10" records, and 10" (or larger-sized) lacquers are used for 7" records. You might be wondering what that extra space is for. This space is needed to properly handle the disc without touching any part of the usable surface. Another is that you want your cutting surface to be as flat as possible, so the farther the imperfect edges are from the music, the better. The aluminum discs get coated in a layer of nitrocellulose, a highly flammable liquid. When it cures, the coated disc has the sheen and consistency of nail polish. The finished disc must be completely free of even the slightest imperfection with a perfect mirror-like appearance.

Next, our lacquer discs head off to the vinyl cutter/mastering engineer for the lacquer-cutting stage. The engineer cuts grooves on only one side of the lacquer disc, one for each side of the record, which requires its own physical stamper when being pressed (typically an A side and a B side). Lacquer masters are made using a disc-cutting lathe. An analog of the source audio is cut into the lacquer with a sharp gemstone stylus that transmits audio vibrations into the surface of the lacquer disc. The lathe is responsible for one of the most complex, crucial, yet misunderstood steps in the record-making process. A simple explanation for how it works is that audio is cued up from either an analog or digital source to feed the lathe. If needed, the audio is processed through equalizers and/or compressors to control the resulting geometric shape of the resulting groove. An amplifier feeds the processed audio to a cutter head, which then passes the signal through miniature left and right side drivers (think tiny headphones the size of your pinky finger) to the cutting stylus that cuts the vibrations as a single spiral groove on the lacquer. A typical groove has a width close to that of a human hair and never overlaps (that would create a skip) until the end of the record, where the groove "locks" in one complete circle. If that explanation still seems unclear, you aren't alone; this is when intangible sounds transform into a written and, more importantly, readable language.

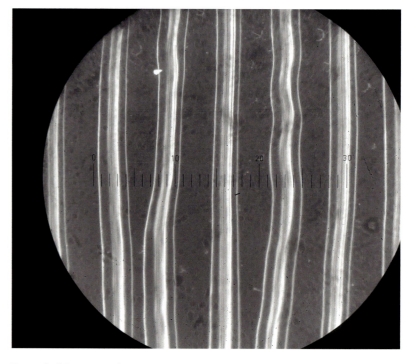

Figure C of Grooves Under a Microscope

The Recording Industry Association of America (RIAA) has defined industry standards or guidelines for cutting lacquers. These specifications seek to improve sound quality and reduce the potential for groove damage that could otherwise arise during repeated playback of your favorite record. During cutting, the engineer adds breaks between the songs to act as helpful visual placeholders by cutting wider spirals between songs so you can easily navigate through the track listing. It's important to note that the final audio mix, equalization, and track order selection must be complete before delivering your music to the mastering engineer for cutting. Not all mastering engineers cover all of these steps of the process.

After the lacquer masters are cut, the engineer focuses on scribing identifiers to the deadwax. They add a catalog or matrix number and a side designator (such as A, B, C, or D) into the part of the lacquer closest to the label. Because it contains no sound, the area near the **center label** is known as "**deadwax**." Grab a record and take a look; you'll find strange symbols, initials by your favorite mastering engineers, or even messages requested by the band or artist. This alphanumeric identifier is especially crucial because it prevents your master from getting lost throughout the entire manufacturing process.

At this point in the process, a lacquer's surface is soft and easily subject to damage. At this stage, you might hear some of these terms: **reference lacquer**, **acetate**, **test acetate**, **dubplate**, or **transcription disc**. Essentially, all of these terms refer to a lacquer that has not been electroplated or "plated." Cut lacquer discs are playable on a regular turntable despite their fragile state. However, playback at this stage is usually to confirm that the project is ready to head to the next step and is not ideal over the long term. Without plating, a lacquer disc has a short shelf life before it starts to degrade. There are somewhat widely different opinions on the maximum amount of time a lacquer can sit before plating, ranging from as little as 24–48 hours to about a month. Storing lacquer discs in refrigerators can significantly extend their shelf life. Still, it is best to quickly move lacquers over to the electroplating process.

It is important to take a moment and discuss an alternate option for creating disc masters outside of lacquer cutting, a process called **Direct Metal Mastering (DMM)**. This process bypasses lacquer discs, relying on a copper-coated disc instead. There are various pros and cons of this process, and people tend to have strong opinions about it one way or the other. Overall, it's essential to consider the final product when weighing the differences between DMM and lacquer-cutting processes.

Some genres, like electronic music, lend themselves well to DMM. In contrast, lacquer-cut pressings have a warmth of tone that can better suit acoustic or human-made music. Sometimes, DMM pressings are described as harsh or sterile, a byproduct attributed to its hard metal surface instead of soft lacquer. Alternatively, some say DMM may reduce noise, creating a clearer sound. Theoretically, lacquers have a much higher risk of accumulating small dust particles on the surface prior to being sprayed with silver during the first stage of the plating process. DMM skips this step by working directly with a copper disc, eliminating these potential extraneous noises from the final product.

The process of how engineers cut audio into both discs is the same; the lathe cuts the audio directly into the copper disc. The rigidity of the copper surface means these discs can go right to plating, skipping the first silver spray in the electroplating process. However, DMM lathes are no longer in production, and current estimates show only about 10–12 currently in use worldwide. DMM is more commonly available as a production technique by plants in Europe. None of the presently accounted-for lathes are operating in the United States for public use. You'll sometimes see a notifier on the cover when a record is cut directly to metal, such as the "DMM" logo or "Mastered on Copper." Keep your eye out and see if you can hear the difference in the sound.

Jett Galindo | The Bakery Mastering
Mastering Engineer
Los Angeles, CA USA

Jett Galindo is a Los Angeles-based mastering engineer and vinyl cutter at The Bakery. Jett has worked on albums spanning a wide array of titles and artists, such as Barbra Streisand, Elvis Costello, Green Day, and League of Legends: Arcane OST, to name a few.

HOW DID YOU GET INTO YOUR INDUSTRY AND WHAT MOTIVATED YOU TO GET INTO IT?

I've always loved music. I lived and breathed it growing up thanks to my parents, who are musicians in their own right, teaching and managing pop/rock bands in the Philippines since I was born in the 80s. The simplest way to explain my motivation for pursuing a career in the music industry is that I'm my most authentic self. Only through this career path have I found my genuine sense of purpose and happiness. I'm grateful that my parents support that; however, they insisted I first get a "real" college degree, and then I could do whatever I wanted.

As a result, I have a college degree in Psychology. Then I went straight to working in a commercial recording facility in the Philippines to prepare for formal audio engineering studies at Berklee College of Music (Boston, MA). After graduating, I interned at Avatar Studios in New York (now known as Power Station). After my 3-month internship, I got hired as the full-time recording engineer to Producer Jerry Barnes (Nile Rodgers, Chaka Khan, Roberta Flack). I thought I was meant to stay in New York until I got shortlisted for a position to be Doug Sax's right-hand woman at The Mastering Lab (Ojai, CA). When I was chosen for the job, I packed my bags and flew to the west coast to work with Doug and the rest of the team. When my mentor, Doug, passed away in 2015, my colleagues from The Mastering Lab and I opened The Bakery in Los Angeles, CA. I've been mastering and cutting vinyl masters here at The Bakery ever since.

WHAT'S YOUR FAVORITE THING ABOUT YOUR JOB AND WHAT MADE YOU FALL IN LOVE WITH WHAT YOU DO?

I'm grateful every day to work with talented musicians, songwriters, producers, and engineers to help them get their music to the finish line, whether digitally or through vinyl. There's never a dull day at work, and I get a lot of joy just hearing every song play through our speakers. It's never lost on me the privilege of getting to do what I do on a daily basis, especially having come from thousands of miles away in the Philippines to live my dream job here in Los Angeles.

WHAT ADVICE DO YOU HAVE FOR SOMEONE WANTING TO PURSUE THIS CAREER?

Pursuing this career isn't easy, especially for women and other minorities. But I urge you to find joy and happiness within the journey itself.

> "Celebrate every win, and don't lose sight of your value and worth. It's tempting to get laser-focused on the destination, but doing so can inadvertently cause you to wear blinders and prevent you not just from appreciating your journey but also from receiving opportunities that may unknowingly be bigger/better for you.

I didn't start my journey knowing I'd be a mastering engineer and vinyl cutter one day, but my path led me to this career, and I'm eternally humbled and grateful for it. It's also important to hang on to your community of like-minded folks as you go along your journey. The individualistic survival mindset isn't sustainable or beneficial to your mental health in the long term.

Allow yourself to thrive within the supportive energy of your peers. I'm very grateful to communities like SoundGirls and Women in Vinyl for allowing me to meet other folks who constantly inspire and support me through their journeys. I also recommend learning about the trailblazers in your industry, especially those with whom you personally resonate. Learning about the women who've thrived in audio engineering motivated me to trust myself and keep going on my path. Shoutout to my personal heroes: Karrie Keyes, Darcy Proper, Mandy Parnell, Aji Manalo, to name a few.

WHAT IS SOMETHING YOU WISH MORE PEOPLE KNEW ABOUT THE WORK YOU DO?

I wish more people knew what it truly takes to make a great-sounding master. There is a misconception that a song/album is magically transformed into a great-sounding record the moment it passes through a mastering engineer. Like our work is akin to the sprinkling of fairy dust. But I think I speak on behalf of all mastering engineers when I say that a big part of a successful mastering session is the communication between the artist, producer, and engineer. Don't hesitate to share your thoughts and feedback throughout the mastering process. Music is a collaborative process, even down at the mastering stage.

Amy Dragon | Telegraph Mastering
Mastering Engineer
Portland, OR USA

Amy, a mastering engineer, prepares audio for digital streaming, CD, cassette, film soundtracks, and vinyl. Mastering at its most basic level is optimizing a release for its distribution format. Her commitment as a mastering engineer is to sweat all the details so that a release can really shine to its fullest potential.

HOW DID YOU GET INTO YOUR INDUSTRY AND WHAT MOTIVATED YOU TO GET INTO IT?

I took an apprenticeship route into this field by way of a studio assistantship. I had no formal audio engineering experience prior, but I had spent 12 years in classical piano training, moving on to orchestral percussion before leaving music performance behind. It turned out that my early foundation in music theory, coupled with a voracious interest in all genres of music, gave me a terrific foundation for mastering. Starting in 2013, under the guidance of my mentor Adam Gonsalves, I spent roughly two years in the studio observing and practicing, reading college audio textbooks, visiting other studios, and pressing plants. Eventually, I worked as a manager at a local pressing plant, Cascade Record Pressing, while learning to cut lacquers and hone my mastering skills. Because I am an independent contractor, mastering was a side gig until I built up enough clients to start mastering full-time in 2018, and I haven't looked back.

WHAT'S YOUR FAVORITE THING ABOUT YOUR JOB AND WHAT MADE YOU FALL IN LOVE WITH WHAT YOU DO?

That moment when a master comes together, and I am deeply moved by the effects of the optimization. As a music-obsessed person, there are few things better in life than that feeling of intimate connection with a piece of music; I get to experience this at my job nearly daily.

WHAT ADVICE DO YOU HAVE FOR SOMEONE WANTING TO PURSUE THIS CAREER?

> " Go for it, but get quality training. There are various paths into this industry. Take 'em if they are offered; seek them out if they are not, but be ready to hustle.

Put yourself in front of quality engineers to learn as much as you can, take college courses, and find a studio or engineer that needs help, but make sure you are taking the time to train your ear for mastering. Quality mastering is primarily about the human behind the board (your judgment, taste, and experience) but also heavily impacted by the room (free from audio reflections to ensure you are correctly editing a frequency that needs it and not just responding to an acoustic anomaly in your room), and your equipment (quality, well maintained, reliable).

WHAT IS SOMETHING YOU WISH MORE PEOPLE KNEW ABOUT THE WORK YOU DO?

Mastering for most formats is pretty straightforward but working on a vinyl release is such a nuanced process. It's the one delivery format where I am not the final stop. There are additional processing steps, multiple quality control evaluations, and human handling that impact the success of a vinyl record. Engineers cutting lacquers bear a great deal of responsibility in preparing a vinyl master, but we are the starting point of a large team of people involved in manufacturing and releasing a vinyl record. From shipping, quality control at electroforming, testing, pressing, and assembly at the plant, as well as distribution practices, it takes everyone's A-game at each step to get a quality vinyl record out to the public.

Heba Kadry
Mastering Engineer
Brooklyn, NY USA

Heba is a mastering and vinyl-cutting engineer based in Brooklyn, NY. She has worked with artists including Björk, Slowdive, Deerhunter, Beach House, Hayley Williams, Animal Collective, Japanese Breakfast, and The Mars Volta, to name a few. Additionally she has mastered original soundtracks for films such as "The Northman," Oscar-nominated for best original score of 2017 "Jackie," "Midsommar," and "The Lighthouse." Heba has also mixed full-length albums, most recently Björk's "Fossora," Jenny Hval's "Classic Objects," and Julianna Barwick's "Healing Is A Miracle."

HOW DID YOU GET INTO YOUR INDUSTRY AND WHAT MOTIVATED YOU TO GET INTO IT?

I was born and raised in Egypt; there were no audio schools or avenues to suggest audio engineering as a future career. However, I was deeply passionate about music, studying classical piano for many years, as well as diving into indie, electronic, shoegaze, and noise in my teens. Still, I didn't think music was something I considered until I started working at an advertising agency composing jingles for ads. I recorded my compositions in studios in downtown Cairo. Instantly when I walked into a control room for the first time, I knew I wanted to become an engineer. Eventually, I traveled to the US to study audio engineering at a short program called The Recording Workshop, and I never left!

I got into mastering because when I was in my 20s, I didn't feel like I was very good as a tracking or mixing engineer. Mastering seemed like the last frontier that I didn't know. I attended the Tape Op audio conference in New Orleans in 2005, and there was a panel featuring mastering engineers from New York. It seemed worth investigating. I was running out of options and money; it was a shot in the dark. I moved to NYC to see if there was room for me to get into this elusive field. If it didn't work out, I would go back to Egypt and compose jingles. Thankfully it didn't come to that! I stuck it out and kept chipping away, working at studios and eventually building my space in Brooklyn. Music has always motivated me, and more importantly, the amazing community and culture it creates. I wouldn't be here if it weren't for the underground indie music community.

WHAT'S YOUR FAVORITE THING ABOUT YOUR JOB AND WHAT MADE YOU FALL IN LOVE WITH WHAT YOU DO?

My favorite thing about my job is the joy of getting lost in your work, where time melts away.

> **"I love being exposed to so many musical genres from all over the world. It's exhilarating and almost like you're crate-digging. I'm a music fan and love the people I work with, our interactions, and the sense of community. It makes me strive to do the best job.**

The privilege of experiencing so much wonderful music from incredible artists always reminds me how super-lucky I am. I do not take it for granted for a second; even on my grumpiest days, I walk into my studio, and all the BS disappears.

WHAT ADVICE DO YOU HAVE FOR SOMEONE WANTING TO PURSUE THIS CAREER?

Get immersed in the wealth of information about mastering and audio online. I wish that existed when I started; it's such an incredible time for knowledge and skill sharing. There are so many incredible YouTube channels now; Sonicscoop, Women's Audio Mission, Dan Worrall, iZotope, to name a few.

Get yourself a mastering **Digital Audio Workstation (DAW)**; there are some affordable ones like Wavelab, Samplitude, and Reaper; get a decent pair of headphones or monitors, treat your room a little bit, and get started. If you're an artist and you've had something professionally mastered, try to reverse-engineer it. There is something beautiful and unexpected to learn referencing those masters or other commercially released albums. It helps develop your listening skills. Reference, reference, reference all the time, and start developing critical learning skills.

We live in an era where anyone can get started in this field, where previously, it was heavily gate-kept by the expense. Shadowing and studio internships are one way to get into it, especially for anyone who wants to learn about cutting lacquers. However, the first step is absorbing all our online knowledge and getting a small system going. You don't need to spend a lot, and it will get you ahead of the curve. You can consider an audio production/engineering school to get a baseline knowledge about audio and its technical theories, but those schools can be cost-prohibitive. I would opt for a short, affordable program because all the learning is from the application or on the job, and it never stops. I learn something new every day.

WHAT IS SOMETHING YOU WISH MORE PEOPLE KNEW ABOUT THE WORK YOU DO?

I wish people were aware of how creative this field can be. There is this fallacy that the mastering engineer is an untouchable wizard you send files to, and they magically come back louder, but there is so much more. Mastering as an art has changed drastically; we have moved away from being the lonesome technician in the back room to a crucial aspect of the production process. We are taught as mastering engineers to bring that last 5%, but if the mixes have issues, I work hard to problem-solve. That's my job; problem-solving and communicating. I hope more people can appreciate how vital mastering is and treat it as a creative aspect of the record production wheel. I'm proud of how much time and dedication I put into my work, as are other mastering engineers I know.

Margaret Luthar | Dark Sky Mastering & NPR
Mastering & Audio Engineer
Los Angeles, CA USA

Margaret is a mastering and audio engineer in both the music and broadcast world. Her career started in classical music recording but veered toward mastering about a decade ago. In 2017, Margaret decided to commit to mastering full-time. At that time, she learned to cut lacquers at Chicago Mastering Service, where she worked as a mastering engineer; later, she took on the role as the head mastering engineer at Welcome to 1979. She is now a freelance mastering engineer in Los Angeles, partnering with Ian Sefchick on their joint venture, Dark Sky Mastering. Margaret also works as an audio engineer for NPR, where she masters podcasts, records interviews, and does location sound work.

HOW DID YOU GET INTO YOUR INDUSTRY AND WHAT MOTIVATED YOU TO GET INTO IT?

I didn't always want to be an audio engineer; as a kid, I wanted to be an actress, a lawyer, a CIA agent, and a musician. In college, I learned that I make music just fine, but I am more talented (humbly so) behind the glass. Thanks to a few classes on studio recording, I got a Bachelor of Music in Music Industry from Syracuse University. I then attended grad school at NYU because I wanted to learn more, and I needed to figure out where my newfound curiosity would take me. A Ph.D.? A studio engineer? Research? I let myself explore a lot in my 20s. I was landing on classical recording but leaning towards mastering. I enjoyed mixing too, but more and more of my work was mastering, and I was starting to enjoy it. I lived abroad for a few years in Norway and, when I returned, decided to explore the mastering side of my life fully. I got a job at Chicago Mastering Service, where I learned to cut and grew as a mastering engineer by leaps and bounds. When I took a job at Welcome to 1979, I stepped up to head of a department, with all the responsibilities of that sort of gig. My career trajectory is different than I expected, but in terms of longevity, it's in a better place. I like that I can build up my mastering base, or if I want to start recording and mixing music again, I can do that too!

WHAT'S YOUR FAVORITE THING ABOUT YOUR JOB AND WHAT MADE YOU FALL IN LOVE WITH WHAT YOU DO?

There's so much! I'll stick to mastering vs. NPR (although I love the variety in that job as well). My favorite thing is the music! A bit cliche, but it's true. I like listening to new music and love pretty much every genre. What's cool about mastering is working with different genres, one week Americana, one rock, etc. The variety keeps it interesting.

In terms of vinyl, being a part of the lineage of cutting, understanding how a record gets made, and being a part of history, I am very fortunate to be a lacquer-cutting engineer. It's specialized and despite all the trials and tribulations of cutting a record, standing on the shoulders of giants, and passing it along to the next generation is extraordinary.

WHAT ADVICE DO YOU HAVE FOR SOMEONE WANTING TO PURSUE THIS CAREER?

Make sure you have a backup plan. If you want to get into audio engineering, get a degree in electrical engineering or computer science or a related field, and music. I've had to teach myself a lot about the electrical engineering side of things. I've been incredibly lucky to be surrounded by very talented technical engineers who have taught me a ton. But I wish I had gone to school for it. Find mentors to help you navigate your interests. It's so helpful to talk to someone who has been in your shoes or at least understands the sometimes volatile world of audio and music. I have mentors from when I was younger that I've circled back to at pivotal points in my life.

> Be open to change, remember to set reasonable boundaries, and assert your needs. Burnout is real; I've experienced it; we all do, even if we love what we do. What matters is how you adapt and change during and after.

WHAT IS SOMETHING YOU WISH MORE PEOPLE KNEW ABOUT THE WORK YOU DO?

Vinyl is complicated. People are complicated. Music is complicated. Combine all three, and sometimes you have a perfect storm! But if you work with good people who love what they do, they'll do their best to make the final product, no matter what format, the best it can be. The same goes for my NPR audio work. It's not just pushing buttons; it's a lot of things all at once. But we're all trying our best!

Mandy Parnell | Black Saloon Studios
Senior Mastering Engineer
London, UK

Mandy is an audio mastering engineer who works with various rare and not-so-rare boutique audio equipment. She has worked on projects with various artists, including Aphex Twin, The XX, Feist, Sigur Ros, Björk, The Knife, Frightened Rabbit, and Brian Eno. She has the exciting role of listening to music all day and "twiddling knobs."

HOW DID YOU GET INTO YOUR INDUSTRY AND WHAT MOTIVATED YOU TO GET INTO IT?

When I was 16, I visited my best friend Julie, who worked at The Manor Residential Studio in Oxfordshire. Just before I left to take the train back to London, the assistant engineer took me into the recording studio, EUREKA! I returned home and looked into courses. At the time, there were only three music production courses in the UK, and I ended up at the School of Audio Engineering. It was their 2nd year of being open. My teachers had written the courses. I dropped out of school at 13 and worked in multiple industries. With each new role, I'd think, "How can I do this for years? I am bored and have only been here for a few months." That eureka moment I followed managed to captivate my attention for all these years. I still love and am fascinated by the exploration of art and science.

WHAT'S YOUR FAVORITE THING ABOUT YOUR JOB AND WHAT MADE YOU FALL IN LOVE WITH WHAT YOU DO?

I work with incredible artists on their art, and I am part of the team working to bring that art to the public. I have loved music since I was a young child. It has been my lifelong friend, there with me in my happy times and saved my life in my darkest times; we have had many adventures together. I have been and still am fascinated with vinyl records. The vinyl mastering process is my favorite part, cutting the master lacquer to be sent to the pressing plant, hearing the test pressings, and putting the needle in the groove. Then, wow, the goosebumps that run over my body when I hear the music for the first time on vinyl. Love!

WHAT ADVICE DO YOU HAVE FOR SOMEONE WANTING TO PURSUE THIS CAREER?

My mother brought me up, quoting, "*nothing's impossible; miracles take a little longer*"; when I found myself frustrated, not understanding something, or finding a task impossible, there she would be. After I graduated, it took me three years to get a paid opportunity in a studio. I knocked on so many doors. I would get disheartened, then play a record I loved, and off I would go again.

I wanted a position at Trident Studios in London, but sadly, they had to shut their doors. I thought that dream was gone. Then I met one of my favorite mentors and teachers, Ray Staff, walking into The Exchange mastering. He had been a mastering engineer at Trident Studios and cut many of my favorite records. I often tell students to look at the producers, engineers, and mastering engineers. You may see a pattern in the names; start there. Assisting one of your favorite producers or engineers is the quickest and most fun way to start learning.

> Where do you envision your career in 15 years, don't waste your time; when you stop learning, do something about it. We are scientists and need to explore.

What helped me was my enthusiasm when talking about music and equipment. If you want to do it and have nothing but sheer excitement and passion for the art and science of audio and a drive to learn, don't quit! It may take time to get there, but nothing's impossible.

WHAT IS SOMETHING YOU WISH MORE PEOPLE KNEW ABOUT THE WORK YOU DO?

An understanding of how hard our job is today compared to when I entered the industry. When I started, we transferred the music from reel-to-reel tapes to vinyl for the consumer. The radio also played vinyl; we had one format. Cassettes were a convenience, not for optimum sound. Now, we have multiple formats and platforms that stream different codecs and high-resolution file formats, plus CDs, vinyl, and cassettes—lots of points to check and many places for error. When I started, the mastering engineer would quality control the test pressings. They would be sent to us from the factory, numbered off the press, to enable us to make an educated view of the vinyl. It is often not the mastering engineer's fault when there are problems, and the misinformation that goes around can waste so much time. I would love more producers, engineers, A&R people, and managers who represent artists to learn more about quality control for vinyl.

CHAPTER TWO

ELECTROPLATING

Once the lacquer is "cut," it's ready for plating. Electroplating is the second step of vinyl record manufacturing. It allows for records to be mass-produced by creating multiple sets of metal "stampers," which have ridges on them and are placed in the record press at pressing plants to "stamp" out copies of the records. The process of **electroplating** coats a metal object with another metal using an electrical current passed through a chemical solution. When a lacquer disc is "plated," we get a metal-coated surface with an exact copy of the playable audio. Think of the lacquer as a template for the electroplating process. The metal copy is more durable and longer-lasting than its lacquer counterpart and is what will be referred to as the "**stamper**." Electroplating is the umbrella term; more precisely, we are "**electroforming**," which is the process used to create stampers. The distinction is that "electroplating" coats a layer onto another metal, whereas "electroforming" creates a separate part.

The first step is to get our lacquer extremely clean by putting it into a series of baths to ensure there is no dust or other particulates. Once prepped, it is ready for **silvering**. The lacquer is affixed in an upright position and sprayed with a very thin "primer coat" of liquid silver for 8–10 seconds, evenly coating the groove-cut surface. This should be done within 48 hours of the lacquer being cut. The next step is called **pre-plate**, where the silvered lacquer is briefly placed into an electroplating bath which coats the silver with a thin layer of nickel. This plating bath operates at a lower temperature and lower amperage to help acclimate the lacquer to the electroplating process. Pre-plating helps eliminate a phenomenon called "**pre-echo**" where an echo or "print-through" in a recording, mechanically induced by a manufacturing fault, is heard before the sound causing it when the recording is played. During the lacquer-cutting process, if the stylus increases too much in temperature, it subsequently increases the surface temperature of the lacquer; the stylus can inadvertently copy a loud signal onto the previous quiet groove, causing an effect known as pre-echo.

The lacquer, which is now silvered and pre-plated, is placed on a round "work holder" that fits into the high-speed rotary electroplating station. The electroplating station includes a "bath" of a green liquid solution called nickel sulfamate filled with nickel molecules from dissolving nickel pellets held in titanium baskets. This solution acts as a conduit for nickel molecules to travel to the lacquer surface, guided by an electrical current. **Rectifiers** located in the bath of the workstation are programmed to deliver specific amperage of direct current, which leads the nickel molecules to the surface of the lacquer. After a little more than an hour, the nickel bonds to the silver, and a nickel plate grows on the surface of the lacquer, creating the "stamper." Other chemicals, including alkaline solutions and cyanide (yes, you read that right), are used throughout the process.

Image of the Electroplating Stations

The next step of the process depends on how many records are ordered for the particular matrix/catalog number and can be approached using one of the following three methods:

One-step process

Two-step process

Three-step process

ONE-STEP PROCESS

This process begins and ends right after the first nickel plate is created. The plate is pulled from our lacquered disc, revealing the newly formed "stamper." The new nickel plate is a reverse copy of the original lacquer cut, meaning its grooves are **embossed** vs. **debossed**. The lacquer is set aside, not to be used again. At this point, the "stamper" is the final product because it can be placed in the record press and used to stamp records. The major drawback of this approach is that there is no master copy for creating new stampers, which can break, get scratched, or get otherwise damaged during use. In a "**one-step**" press run, if the "stamper" is broken or scratched, you would have to go back and cut a new lacquer and start over, which can be costly and time-consuming. It is not recommended for more than 500 copies.

TWO-STEP PROCESS

Alternatively, the "**two-step**" process creates a master copy called a "**mother plate**" that can be electroplated multiple times to develop additional "stampers." Like the one-step option, the silvered and pre-plated lacquer is placed into a high-speed rotary electroplating station. After a little more than an hour, the "**father plate**," or "father stamper," is created, and the lacquer is set aside, not to be used again. The "father stamper" is then de-silvered and placed back into the high-speed rotary electroplating station, and after 2 hours, it produces a thicker metal copy called a "mother." The mother is logged and carefully stored to be used later if another stamper is needed. This "father stamper" is the final product and can now go into the record press as a stamper to press records leaving a backup copy, the "mother," to produce additional stampers. One "mother" can usually create about ten stampers, and one stamper can press about 1,000 records before it wears out. For many albums, 10,000 copies are more than sufficient for their release.

Because the grooves on a "mother" are like they are on a record, it is possible to play them on a record player. You don't need a special stylus to play them; however, you do need to replace the stylus regularly as the nickel wears heavily on it. The only consideration is that not all turntables can accommodate a 14" mother. Usually, a spin on a record player acts as a quality control check before the "mother" creates new stampers.

THREE-STEP PROCESS

The final option, called the "**three-step**" process, is typically used for production runs larger than 10,000 copies or that need several represses. In this case, the silvered, pre-plated lacquer is placed into the high-speed rotary electroplating station for 2 hours, which produces a thicker "father stamper" than in the one-step process. The lacquer is set aside, not to be used again. The thicker "father" is then placed back into the electroplating station and, after another 2 hours, produces an equally thick "mother." The "father" is logged and carefully stored to be used if another "mother" is needed. The "mother" is placed back into the electroplating station and, after a little more than an hour, produces a thinner "stamper" for pressing. The stamper is, again, the final product. Now, you have safety copies for producing both of the most critical parts in the pressing process. After the "mother" has generated her ten stampers, it is retired. Then the original "father" comes out of safekeeping, and the electroplating process repeats. A new mother and more stampers are possible without returning to a new lacquer cut.

Each side of an album has a stamper plate, an A-side stamper, and a B-side stamper. In each of the previously mentioned processes, the "stampers" are finished by having the back (non-grooved sides) sanded smooth to ensure they are of equal thickness and to eliminate surface anomalies. Any slight imperfections accumulated at any stage in this process can transfer to the record.

The final step is to punch out the center hole for the mould pin using a microscopic scope. The scope uses the grooves as reference points to find the stamper's true center, which ensures the stamper fits perfectly into the press. Next, a forming machine will shape the stamper to fit a particular press. From there, you're ready to go into production!

If DMM piqued your interest in the last chapter, the electroplating required to produce stampers for them requires fewer steps. As opposed to one, two, or three steps needed for conventional lacquer masters, copper master discs can withstand repeated trips through the electroplating process to produce the necessary number of stampers using the one-step plating process and retain their integrity. The copper plate acts as the "mother" to create all the stampers.

ELECTROPLATING PROCESS FOR VINYL

1 STEP PROCESS
CAN PRESS ~1,000 RECORDS*

2 STEP PROCESS
CAN PRESS ~10,000 RECORDS*

3 STEP PROCESS
CAN PRESS ~100,000 RECORDS*

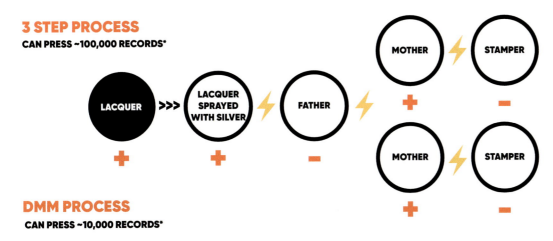

DMM PROCESS
CAN PRESS ~10,000 RECORDS*

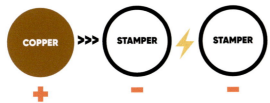

⚡ = ELECTROPLATING BATH

✱ = 1 STAMPER CAN PRESS APPROXIMATELY 1,000 RECORDS AND EACH MOTHER CAN MAKE ABOUT 10 STAMPERS

Yoli Mara | Welcome to 1979

Owner

Nashville, TN USA

Yoli Mara is the co-owner of several closely related businesses, all in the music industry. She and her husband have owned a recording studio, Welcome to 1979, since 2008. They have a large facility in Nashville, Tennessee, that started out offering tracking and mixing. In 2013, they made their first foray into the vinyl side of things with the purchase of a Neumann VMS-70 lathe and began offering vinyl mastering and lacquer-cutting services. A few years later, they founded Welcome to 1979 Industries, an electroplating facility where they manufacture stampers for vinyl production.

HOW DID YOU GET INTO YOUR INDUSTRY AND WHAT MOTIVATED YOU TO GET INTO IT?

After owning the recording studio for several years, we were excited by the vinyl resurgence. This led to the addition of the cutting lathe. Once we'd been cutting lacquers for a while, we saw a bottleneck at the electroplating stage of the process and a real opportunity to offer lacquer cutting and plating in the same facility. We are able to offer a level of service that other facilities cannot.

WHAT'S YOUR FAVORITE THING ABOUT YOUR JOB AND WHAT MADE YOU FALL IN LOVE WITH WHAT YOU DO?

My favorite part of my job is being able to be a part of so many different types of projects. Every single project that is going to be released on vinyl needs to be cut and plated, so we are involved in projects of all sizes, from indie artists pressing 50 copies to major label artists pressing tens of thousands. To projects in all genres, from Americana to rock, to stand-up comedy, to country, to death metal, to silent grooves for meditation. It's never the same thing and always fun to pop something on for Quality Control (QC) and not have any idea what you're going to hear.

WHAT ADVICE DO YOU HAVE FOR SOMEONE WANTING TO PURSUE THIS CAREER?

> " Do your homework! There is not a single step in the vinyl manufacturing process that is simple, but there are tons of experts out there willing to open their doors to you and share their knowledge.

Attend conferences like Making vinyl (they have a lot of useful information on their website as well), and reach out to the newly established trade organization, the Vinyl Record Manufacturing Association.

WHAT IS SOMETHING YOU WISH MORE PEOPLE KNEW ABOUT THE WORK YOU DO?

I wish people understood how much went into getting a project released on vinyl. It is a lot more time-consuming than a digital release, and if more folks understood the process, they would be better prepared and less surprised by the lead times and the number of moving parts involved.

Janine Lettmann | Pallas GmbH
Electroplating QC "Mutterstecherin"
Diepholz, Germany

Janine has been working for the German pressing plant Pallas for over ten years. Pallas is a family-run business that has been producing vinyl records since 1948. Her position in quality control of the electroplating department is called "Mutterstecherin," which translates into "mother stabber," as funny in English as in German. After Janine's colleagues in the electroplating department turn the lacquers into father and mother plates, Janine's job begins. Electroplating is a highly mechanical process; small nickel particles can get stuck in the grooves leading to static noises such as clicks, pops, and rattling sounds. Upon detecting any flaws, she corrects them using a tiny graver on grooves magnified 200 times under a microscope. Needless to say, Janine has a calm, steady hand not to damage the grooves.

HOW DID YOU GET INTO YOUR INDUSTRY AND WHAT MOTIVATED YOU TO GET INTO IT?

After I finished school, I was looking for a training position, and Pallas was searching for industrial clerk trainees back then. I am a big music enthusiast, so training at Pallas was my first choice. Luckily, it was a match. After finishing my traineeship, I worked as a shipping manager preparing customs documents and such. Even though I was having a lot of fun in my old department, I didn't have to think twice when I was offered a position in QC of the electroplating department. I mean, what's better for a music nerd than listening to music all day long? I was also super excited to be more involved in the creation of the product itself.

WHAT'S YOUR FAVORITE THING ABOUT YOUR JOB AND WHAT MADE YOU FALL IN LOVE WITH WHAT YOU DO?

I have always been drawn to music, so it is very exciting for me to hold a job in the industry, especially since we are located in the country. The coolest thing about this particular job is that I can actually secure our production with my work. If something goes wrong during a press run, we don't have to request new lacquers from the cutting studio right away because I am able to repair the metal if needed. Besides that, you never know what the day will bring. Some days, I am one of the first to listen to new releases by top artists playing concerts at the biggest venues you can imagine. On other days I check mothers for titles I've never heard of and discover new stuff that I genuinely enjoy. Not to mention the good old classics kept alive by all the vinyl lovers out there. It's the best of both worlds!

WHAT ADVICE DO YOU HAVE FOR SOMEONE WANTING TO PURSUE THIS CAREER?

Well, if you're interested in this specific job, I can only suggest you move to Diepholz, Germany, and work at Pallas. I really don't know if any other pressing plant offers this position. Even in Germany, it's such a rare job that it has been used as a search term on several TV quiz shows.

> " You should have a good ear, a steady hand, and a strong ability to concentrate. Besides that, being interested in vinyl and music is always a good start.

WHAT IS SOMETHING YOU WISH MORE PEOPLE KNEW ABOUT THE WORK YOU DO?

I am sure many reading this have never heard of this job in the first place. So being able to introduce it on the Women in Vinyl platform has been a huge step for me. I'd like to take the opportunity to raise awareness for all the hard work and love every pressing plant and its employees are putting into records. Even though the demand has never been higher, I still don't consider vinyl records as fast mass-produced products like CDs, for example. There are still actual human beings like me and my colleagues touching every single disc we make, putting our thoughts into improving this highly analogue technique and carrying on the legacy of this emotionally charged medium. That being said, I really hope we can keep my job alive in the future. My work plays a big role in providing products of exceptional quality, but it also takes years to learn your ways with the graver on this sensitive material. It's a real craft, just like every record itself is one of a kind. I love that.

Desiree Oddi | Record Technology Inc.
Plating Tech
Camarillo, CA USA

Desiree is the Plating Tech at Record Technology Inc. (RTI). Their plating department manufactures masters, mothers, and stampers for records that will be pressed anywhere in the world or right there in their own facility. Her responsibilities mainly center around customer service, coordinating orders, deadlines, and shipping details. In addition, Desiree helps with quality control by visually checking stampers for flaws or defects and occasionally making repairs to metal mothers to eliminate pops or ticks.

HOW DID YOU GET INTO YOUR INDUSTRY AND WHAT MOTIVATED YOU TO GET INTO IT?

I learned about this industry because of my Dad. He has worked in plating since before I was born, and I visited RTI a lot growing up. I began working as a customer service representative in pressing and quickly transitioned to the position in plating.

WHAT'S YOUR FAVORITE THING ABOUT YOUR JOB AND WHAT MADE YOU FALL IN LOVE WITH WHAT YOU DO?

Working in plating gives you an interesting behind-the-scenes look at the music industry. Plating is a very hands-on process. From the time lacquers arrive and a unique ID number is hand-etched into them, to when the metal is completed, my colleagues and I have handled each of these parts and spent a good amount of time inspecting, cleaning, and listening to them. It's really cool to see a record in a retail store and recognize it as one that we plated.

WHAT ADVICE DO YOU HAVE FOR SOMEONE WANTING TO PURSUE THIS CAREER?

I've found that it's very helpful to familiarize yourself with the whole process of vinyl manufacturing, not just plating.

> **„ Having the big picture makes it much easier to determine the cause of any audible or visual defects that may appear on the metal or the vinyl.**

WHAT IS SOMETHING YOU WISH MORE PEOPLE KNEW ABOUT THE WORK YOU DO?

When it comes to vinyl record manufacturing, plating focuses a lot more on chemistry and machinery than the music itself. A great-sounding record is the end goal, but it takes a lot of science to get there.

Elsie Chadwick | Stamper Discs
Office Manager
Sheffield, UK

Elsie is the Office Manager at Stamper Discs, a plating facility serving the vinyl manufacturing industry based in Sheffield, UK. She handles customer service as well as business administration alongside her colleague Sharon. They are the first port of call for clients consisting of mainly pressing plants and record labels, and they are on hand to answer any questions and solve any problems they may have. Elsie ensures all the lacquers and copper parts they receive from the cutting studios are booked in for the correct plating process. Once an order has moved through production, they'll give it a final QC check and ship the stampers off to the customer. Elsie leads production planning to ensure the customer service and production teams are in sync with regard to their customers' projects. She has recently been given the opportunity to be involved with accounting, recruitment, and training.

HOW DID YOU GET INTO YOUR INDUSTRY AND WHAT MOTIVATED YOU TO GET INTO IT?

I applied to work at Stamper Discs after seeing the role advertised online. I had graduated with my master's degree earlier in the year and was working in hospitality. I knew this wasn't something I wanted to do in the long run, but with the covid-19 pandemic in full swing, my options were limited or had at least been put on hold. The idea of working in the music industry has always appealed to me. But at that point in my life, I hadn't given too much thought about how I could do this. Having read up on the vinyl industry, I realized there was a whole world to learn about and get involved in. It looked like a really interesting role within a small and friendly team, and although I knew very little about stamper production, I tried to find out as much as possible and applied. It also helps that the only place in the UK to offer a trade **galvanics** service to the vinyl industry is a 20-minute walk away from my apartment!

WHAT'S YOUR FAVORITE THING ABOUT YOUR JOB AND WHAT MADE YOU FALL IN LOVE WITH WHAT YOU DO?

Everyone at Stamper Discs probably agrees that one of the best parts of the job is being able to listen to, and be involved in, the production of the varied range of projects we receive from our clients. Sometimes this means getting to hear the upcoming album of one of your favorite artists and contributing to its physical release, which has happened to me a few times!

Being on the customer service side of things also means I get to speak with people from all over the world. It's always interesting to learn about what kind of records they're pressing and the sort of projects they're working on. On a personal level, it is also brilliant to be a part of the Stamper Discs team and company, and I'm grateful to be able to work with such a hardworking bunch of people.

WHAT ADVICE DO YOU HAVE FOR SOMEONE WANTING TO PURSUE THIS CAREER?

If you see the opportunity go for it! Even if somewhere isn't recruiting, there's no harm in showing your interest and making initial contact.

> " Don't be put off because you think you lack knowledge or experience in a certain area. It can seem like an intimidating industry to try and get your foot in the door, but everyone has to start somewhere.

I would also advise you to do as much research as possible and ask questions when and where you can.

WHAT IS SOMETHING YOU WISH MORE PEOPLE KNEW ABOUT THE WORK YOU DO?

I wish more people were aware of the plating stage in the vinyl production process and the role of stampers on a record press. More people seem to be aware of vinyl production's cutting and pressing stages but not so much of the galvanics. I remember doing research for my interview and finding it difficult to find detailed resources on the production of stampers compared to other aspects of the industry. It would be good to get more light shining on the precision and detail required at this stage. But I think with more information becoming accessible to a wider audience through collectives such as Women in Vinyl, a growing number of people are discovering the importance of this part of the process.

Emily Skipper | Press On Vinyl
Galvanics Technician
Middlesbrough, UK

Emily, a Galvanics Technician, works at the newly built galvanics lab at Press On Vinyl in Middlesbrough. She works to make stampers for their presses by spraying the lacquers with silver, then running them through the galvanic baths to form a negative version of the lacquer made from nickel. They are then off to coin, form, and clean to make it the right shape stamper for the mould, then it's off to the press.

HOW DID YOU GET INTO YOUR INDUSTRY AND WHAT MOTIVATED YOU TO GET INTO IT?

Before I started at Press On, I had absolutely no idea about the processes required to make records. When I learned about the factory opening, I decided to read about it, and although I did think it sounded quite complicated, it sounded really interesting and right up my street. I was buzzing when they offered me a job but nervous because I hadn't done any work like it before; the best I'd done was some shoddy DIY. It was comforting knowing that I was starting as the factory was starting, and that meant, hopefully, I wasn't the only person who had no clue what they were doing and that we would all learn it together. I've loved learning the job, and I'm very grateful for being given the opportunity. If I'm honest, I got into the industry by being very lucky and knowing great people who have taken a chance on me.

WHAT'S YOUR FAVORITE THING ABOUT YOUR JOB AND WHAT MADE YOU FALL IN LOVE WITH WHAT YOU DO?

My favorite thing about my job has got to be my team; every achievement we've made wouldn't be half as good if they weren't such a great group of people. We spend all day working hard and dealing with all the issues new businesses face, but I still come to work smiling each day because of them.

WHAT ADVICE DO YOU HAVE FOR SOMEONE WANTING TO PURSUE THIS CAREER?

> " My advice is to read and learn as much as you can about the process just from searching the internet. There's loads of helpful sites and forums and even videos showing how people make their own stampers.

Everyone does it slightly differently, but the general idea is the same. I found a really helpful handbook, just available to download for free, about nickel plating, which has helped me figure out a lot about what happens in the galvanic baths. I did find that when I first started reading, I couldn't quite wrap my head around certain parts, but the more you read, the easier it is to understand.

WHAT IS SOMETHING YOU WISH MORE PEOPLE KNEW ABOUT THE WORK YOU DO?

I wish people knew about the effort that goes into producing vinyl. I don't imagine many people outside the industry are aware of the lengthy process, from recording the music to packing the final record and sending it out the door. It has made me recognize just how much of a piece of art a record is; the love and care that goes into every step within our factory is phenomenal.

CHAPTER THREE

MANUFACTURING

Before we jump into the nitty gritty and science of record manufacturing, let's take a moment to review a brief history of how our beloved black disc has evolved. The first vinyl discs are referred to as "78s" because they play at **78 rpm** (**rotations per minute**). These discs are composed of shellac resin which feels a bit heavier and more brittle than today's vinyl. During World War II, shellac was hard to come by, so production began shifting to more available vinyl materials. In 1930 RCA started to make records played at **33 1/3 rpm** and pressed onto 12" diameter flexible plastic discs. These were commercially unsuccessful because consumers needed turntables with a 33 1/3 rpm setting. Not to mention, during the Great Depression, the sale of new machines was minimal. In 1939 however, Columbia Records picked up RCA's idea and continued developing vinyl technology. By 1948, Columbia introduced the 12" **Long Play (LP)** 33 1/3 rpm **microgroove record** created by Peter Goldmark. This vinyl record had a recording capacity of around 21 minutes per side. At the same time, RCA Victor was introducing a competing format, the 7" or **45 rpm Extended Play (EP)**. From here, the rivalry between RCA Victor and Columbia Records began. Between 1948 and 1950, both formats fought for prominence in the market; it was the "War of the Speeds." In the end, the 12" LP, 33 1/3 rpm became the dominant format, and the 7" record became the preferred choice for singles. Singles, 45s, or EPs offer a similar playtime to **78 rpm** discs, about 7 minutes for the best sound. In both cases, the length per side remains consistent today as a general rule. In the early 1960s, consumers finally caught the vinyl bug, and the market shifted again from **Mono** to **Stereo**. Production of Mono LPs faded out, finally ending by 1968.

THE CHEMICALS

Here we are, how the sausage gets made, as they say. The next step is pressing the vinyl record we love so much. The process requires a few crucial chemical components, the first of which is the most critical ingredient, **PVC** or **polyvinyl chloride**. This chemical is the foundation of all your discs. You might be surprised to hear that PVC, in its pure form, is translucent white and somewhat brittle. Special additives help make the end product softer, more pliable, and more colorful. These additives fine-tune the mix for just the right product needed to make vinyl records. The PVC used for vinyl record pressing has more additives than the mix used to make PVC pipes or vinyl flooring. However, the underlying PVC compound is the same.

Let's examine the unique mix used to create our perfect musically inclined blend of PVC. The ideal combination produces a final product strong enough to support the thinly pressed grooves. It also allows the turntable styli to ride along the troughs without damaging the surface. The final composite is also quieter and less brittle than its shellac predecessor.

The first additives are **polyvinyl acetate (PVA)** and **plasticizers**. They are essential in forming the record; they are responsible for improving flexibility and increasing the disc's durability, ultimately reducing breakage. They also create a better-sounding match to the original master lacquer. **Lubricants** are another additive, and during pressing, they help the hot puck of PVC in the machine to spread more consistently across the surface of the stamper during production. On the final record, they also reduce friction, providing a smoother glide of the needle. The PVC mix also gets a dose of **heat stabilizers** that help to make the overall compound more robust and neutralize hydrogen chloride gas. This gas forms while our puck is in the press, which reaches temperatures of about 345°F (160°C). Hydrogen chloride gas can cause other compounds in your record to break down. Thankfully heat stabilizers help guarantee those chemical breakdowns don't occur and ensure your vinyl record remains stable.

The final and most visually noticeable additives are **colorants**! The most notable is **carbon black**, creating the familiar black vinyl discs we all know and love. Carbon black, as a bonus, offers the PVC mix some additional durability. Some parties believe carbon black sounds better. The idea is that it reduces static as it distributes and dissipates electrical charges, thereby reducing the **noise floor** of the recording. Color is a highly debated topic, but most people are unlikely to discern a significant difference between a black or another color vinyl pressing. In truth, many factors are at play when examining how color vinyl sounds. Factors include the manufacturer of the chemical compounds, the pressing machine settings, the environment at the manufacturing facility, and even the machine itself. Initially, most records were black, but today the color options span the rainbow and beyond. If it goes with your overall packaging and you love how it looks, go for it!

Now let's take a moment among all this talk about chemicals to recognize that the record-making process isn't what we call "environmentally friendly." Vinyl records are not biodegradable. They are, however, archival, meaning you can stick them on a shelf, and they will remain intact and playable long into the future. PVC is made from oil and takes up to 1,000 years to decompose in a landfill. There are pressing plants working to offset their carbon footprint and recycle vinyl into other products. Some companies try to reduce waste by grinding-up rejected records and mixing them into the composite of future discs. That process is called regrind. Other companies are looking into ways to make eco-friendly mixes and even revolutionize the manufacturing process with more energy-efficient machines. We look forward to where these incredible innovations take the record-pressing industry in the next ten-plus years!

THE MACHINES

Now, let's move away from the chemicals and discuss the machine that makes all our dreams come true, the **record press**. In simplest terms, it uses heat, steam, and pressure to manufacture vinyl records. Imagine for a moment an automated supersized waffle maker.

Getting these machines ready to press your new favorite album begins with putting the nickel stampers we discussed in Chapter 2 on the press. Stampers are uniquely fitted to the **moulds** on the machines at the pressing plant they are going to. The moulds have channels inside them for steam and cold water to circulate through them in order to heat and cool the PVC, and are what the stampers are placed onto inside the record press. Different pressing plants have different machines, and each machine has a unique mould profile; it is not a one size fits all situation. Step one starts as a press operator places the A side and B side stamper on the moulds and into the press.

Next, our unique blend of PVC in the desired color(s) fills the machine's hopper. Center labels are placed into a label cup and put into the record press. Labels may look innocent enough, but you might be surprised to hear that they require baking in an oven before being utilized on a record press. This visit in hot temperatures dissipates moisture from the paper and inks to prevent them from tearing in the press, or what looks like an exploded label on the record.

Finally, the machine has stampers, PVC, and labels all loaded in. The job can begin! Time to fire up the press, as the press operator dials in the correct settings to make a flat, high-quality record. The appropriate settings can be inconsistent, and the operator masters the art of when and how to adjust for issues as they arise. The press heats the PVC and extrudes it into a cup which forms a vinyl **puck** (or "biscuit") to the correct amount of PVC for the job weight, either 7" ~40g or 12" at ~140g or ~180g. The labels are dispensed on either side of the puck to correspond with the A and B stampers. The stampers are pressed down with about 100 tons of pressure over the labels and puck, filling the grooves in the stampers to form the finished record. As the PVC moves across the stamper, an excess amount spills out the side, like when you make waffles and batter flows out the side. We call the spillover material **flash**. Cool water circulates through the moulds as the pressure between the stampers releases. The press releases the record before trimming the flash from around the outer edge. Once the presses are up and running, each can (theoretically) press about 100 records per hour or about 30–40 seconds per record.

PROCESS AND PRESS INNOVATION

The comeback of vinyl has outpaced the ability of the manufacturers to keep up with demand as of late. They face capacity limitations, supply chain shortages, and soaring demand. The lack of innovation is a big reason the industry continues to contend with some of these challenges. The lag goes back to the 1950s–1960s; the medium's near-extinction in the 1990s maintained stagnation, and a swift resurgence exasperated the situation.

Companies like Hamilton or Toolex Alpha made many record presses that fueled vinyl's golden years. In 1984 the last new record press rolled off the assembly line as the cassette overtook vinyl in sales. When vinyl fell out of fashion, the machines sat idle. In 2017 new pressing plants started to open again using the same old machines. These new owners found themselves outbidding each other to buy used, rusty, sometimes derelict equipment, which needed love and refurbishment.

The situation prompted Canada-based company Viryl Technologies to create the first new record-pressing machine, the WarmTone. Viryl reverse-engineered functionality of the old presses while adding new enhancements like robotics, HMI (Human Machine Interface) systems, and sensors. In 2006 former Toolex Alpha employees in Sweden formed Pheenix Alpha. They revitalized the old Toolex Alpha model's profile while adding the same new technologies and systems like Viryl to their trademarked Toolex machines.

All in all, new machines provide more automation and technology. The automation allows for quicker cycles than a manual or semi-automatic press, which you would use to make a lot of the crazy effects you see on social media. However, the process still relies on compression molding and has mostly stayed the same since the 1950s–1960s. Production still relies on a boiler to create the steam that goes into the moulds and compresses the puck of PVC. Innovations in the manufacturing process, including new machines that rely on injection molding rather than a press, are in development for the future.

FROM PRESS TO PACKAGE

Beyond the pressing process, other things also happen at the pressing plant. Deciding on the right plant for an artist's project is influenced by quantity, product, and packaging. Part of that process is identifying who offers the desired quantity. Due to the process involved, most pressing plants have a minimum of 200–500 pieces for each run on a press, which imposes a significant upfront cost. After choosing a manufacturer, the artist works with a sales representative. Sales representatives guide customers through all the options to create the perfect package for their release. They discuss all the finer details about how the artist wants the final product to look, from packaging to color and weight. Once the order details are complete, the sales rep will guide the customer and project into production.

Record weight, 180g vs. 140g, is another decision and highly debated topic. There is a perceived added value to a 180g record due to the potential for less vibration on the platter during playback. However, any potential difference in sound is generally insignificant for most turntables and their listeners, meaning that the only real difference between the two weights is weight. For more details and discussion on this topic, we recommend checking out the Women in Vinyl, Vinyl 101 podcast episodes (htttps://womeninvinyl.com/podcast).

Once the order is in production, the album artwork is submitted. Unfortunately, the artwork is often an overlooked part of the process and remains outside this discussion's scope. That said, specific vendors typically print the record jackets and other components like inserts, download cards, and posters. A prepress staff member at the pressing plant is a bridge between the printer and the customer. It is their job to oversee correct art placement in templates and file set-up before sending anything to press. While the art is **proofing**, the first records (AKA **test pressings**) are likely also back with the customer for review. Test pressings provide a moment to confirm everything is right before the entire run goes to press. They ensure the audio is correct, that the tracks are in the correct order, and that there are no consistent pops or clicks on all the copies to take you out of the listening experience.

The QA (quality assurance) or QC (quality control) department steps into the process at the critical "test pressing stage" and again during the final run of an order. This individual listens for possible pops and clicks before the test pressing goes to the labels or artists for review and then checks again during the final run. These individuals have trained ears to hear for issues that can occur during production, like **stitching** (when the material has been cooled down too far at the end of the press cycle). They also listen for **nonfill**, when the material going into the puck isn't hot enough, interfering with filling the stampers' grooves, and other audible or cosmetic issues. Pressing technicians cannot fix audio levels, sibilance, and the like at this stage. The QA/QC individuals genuinely listen to ensure the manufacturing process creates an excellent final product and listening experience.

Once the test pressings are approved, the order goes into the production of the finished goods. The records get another round of quality control after they come off the press before heading to the packaging department. In packaging, they get a final visual inspection as they go in dust sleeves. Then each record is slipped into its custom jacket, shrink-wrapped or polybagged, sometimes stickered, and shipped off into the world.

Caren Kelleher | Gold Rush Vinyl
Founder & CEO
Austin, TX USA

Caren moved to Austin in 2017 from the California tech world to start Gold Rush Vinyl, a new record-pressing plant bringing fresh eyes to the industry. Today they employ about a dozen people and also have a business making 24-karat gold records.

HOW DID YOU GET INTO YOUR INDUSTRY AND WHAT MOTIVATED YOU TO GET INTO IT?

My career in vinyl started because I saw both a problem and an opportunity. I had been working in the music industry since college and thought I had seen it all, but I had never seen anything like the vinyl resurgence. I worked at Google, where I helped launch its streaming music service and the Google Play Store. By the end of my tenure there, I was leading Google Play's partnership work with other music app developers like Spotify, Pandora, and Shazam. In my spare time, my sister and I also worked with a few indie bands we loved. That experience as a band manager gave me a unique perspective on the music industry and the work I was doing in Silicon Valley. The most striking observation was how important merch like vinyl had become to musicians in the age of digital music.

One day at work, I was giving a presentation about the state of the music app ecosystem, and I shared an important statistic that inspired Gold Rush Vinyl. That was: for the average American musician to make minimum wage, they can either sell 100 vinyl records or accumulate nearly 2 million streams on YouTube. I remembered my experience trying to get vinyl made for a young band. No pressing plant would take our order, even when I offered to pay rush fees. They were too swamped with existing work to prioritize a new indie band and told me vinyl was a process that couldn't be rushed. That's when I saw the opportunity to start Gold Rush Vinyl as a pressing plant that catered to these artists and helped them earn more money through vinyl. One morning I went to a coffee shop and drew up a business plan for a new record-pressing plant called Gold Rush Vinyl. I took it to my local bank, and they helped me get a meeting at Wells Fargo corporate headquarters in San Francisco to pitch my idea. I walked out of that meeting with an initial funding commitment and called my family to tell them, "I think this means I'm starting a record-pressing plant."

WHAT IS YOUR FAVORITE THING ABOUT YOUR JOB AND WHAT MADE YOU FALL IN LOVE WITH WHAT YOU DO?

The best part of my work is creating jobs, especially for other women. I'm very proud that Gold Rush Vinyl has remained a woman-owned company and now employs a team that is 50% women and growing. I love mentoring college students, too, and helping them see all the possibilities in vinyl and beyond. While a lot of attention is paid to the innovative things we're doing in our business, the high quality of our records is due to the hard work and care that each team member puts into our products. It's a joy to come into the factory each workday, be surrounded by such smart, kind, and clever people, and see how they make the company better with their ideas and contributions.

WHAT ADVICE DO YOU HAVE FOR SOMEONE WANTING TO PURSUE THIS CAREER?

> " Don't sit back and wait for the perfect moment, or the perfect song, or the perfect business plan before you start making moves.

Too many people — especially women — talk themselves out of new career opportunities because they don't feel ready or don't feel like they're good enough. I didn't have any experience in vinyl manufacturing before I founded Gold Rush Vinyl. I did have a solution to a problem, a willingness to learn, and an enthusiasm for the work ahead. That was enough to get started. I've made a lot of mistakes along the way, but I've also created a life that I'm very proud of.

WHAT IS SOMETHING YOU WISH MORE PEOPLE KNEW ABOUT THE WORK YOU DO?

Vinyl manufacturing is extremely hard but also extremely rewarding. Had I known how many things could go wrong (pipes bursting! plastic shortages! raccoons in the boiler room!) I might have been scared off from starting a pressing plant, so I'm glad I came into this job a bit naive. That said, no two records will ever come off the presses the same, and it's important to remember that vinyl is a manufactured product with its own quirks. Vinyl won't sound the same as a digital album, and that's the point. It's part of what makes vinyl so special.

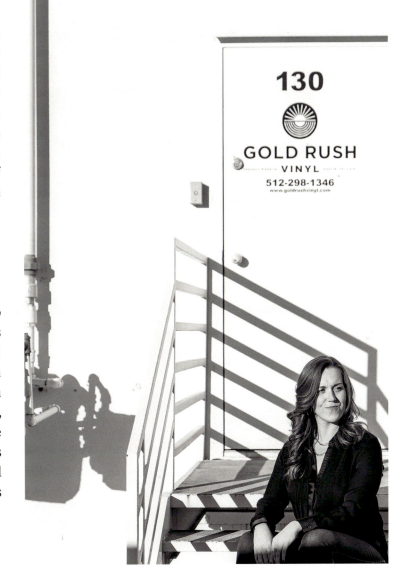

Anouk Rijnders | Record Industry
CCO
Haarlem, The Netherlands

Anouk is the COO of Record Industry, one of Europe's largest vinyl pressing plants based in the Netherlands. Anouk grew up with vinyl, listening to Disney fairy-tale records as a toddler. Later she bought her own records like Kiss, Prince, Duran Duran, and more with her pocket money. One of her first jobs was in a CD shop at the beginning of the CD era, which was unique at the time. After that, she worked at a vinyl shop way before the vinyl resurgence; the rest is history.

HOW DID YOU GET INTO YOUR INDUSTRY AND WHAT MOTIVATED YOU TO GET INTO IT?

I studied at the Art Academy and worked as a producer and director for several television programs. In 2000 my friend, Ton Vermeulen, asked me if I wanted to work for him at a vinyl pressing plant he had taken over from Sony Music two years prior. It was crazy busy at that time, and he needed a troubleshooter. The funny thing is, I had doubts about the job at first. Twenty-three years later, I am still here and enjoying every minute. Each year has been different. From very hectic times to a really slow and uncertain period, and now extremely busy again, pressing 40 to 50,000 records a day, an unthinkable quantity for us 14 years ago! Right before the pandemic started, we thought we were over the peak of the vinyl demand. Who would have thought the pandemic, which has been devastating for so many people in many ways, would lead to the next vinyl resurgence? For us, as a company, it resulted in some major changes. This year we merged with Bertus Distribution, Europe's largest independent distributor of physical products, which also owns a few well-established record labels. As a group, we invested in new vinyl presses, and we're upscaling our capacity from 32 to 62 presses, enabling us to press 25 million records a year.

My job has changed from troubleshooting to sales manager and CCO over the years. Many different aspects of my work make me feel so blessed and keep me motivated. Next to the daily work in the office, I got the chance to produce two books: *Passion for Vinyl Part 1* and *Part II*, and I am working on a new book as we speak. Another great project was building a one-of-a-kind Mastering Studio and Live Room called Artone Studio. Artone offers direct-to-disc cutting and recording on unique vintage equipment. I'm working on a new project called Haarlem Vinyl Festival, an international live music and arts & culture festival, including a conference, all evolving around vinyl. The first edition will be in September 2023.

WHAT'S YOUR FAVORITE THING ABOUT YOUR JOB AND WHAT MADE YOU FALL IN LOVE WITH WHAT YOU DO?

Record Industry runs like a family business, despite our team having over 140 employees now. It's a very social environment where we engage all colleagues and make everyone aware that each position is important in the "vinyl production chain." I like that the product we make has a magic about it. I am still amazed by this old technique; so simple but oh so brilliant. It's that magic and what it does to people, the emotion involved, the listening experience, collecting and admiring the artwork which makes me so passionate about my work.

WHAT ADVICE DO YOU HAVE FOR SOMEONE WANTING TO PURSUE THIS CAREER?

I don't have a "tip." The only thing I can say, which is a personal opinion that can apply to many things, is:

> **"It is important to take pride in what you do, no matter which job or project. It helps if you love what you do, as it makes things easier.**

I am very grateful for the trust and freedom Ton and Mieke Vermeulen (the owners of Record Industry) have given me to pursue my ideas and allow me to do so many great things, which contributed to making Record Industry a brand that stands for a high-quality product but also a company with a passion for the vinyl format in many ways.

WHAT IS SOMETHING YOU WISH MORE PEOPLE KNEW ABOUT THE WORK YOU DO?

It's not so much that I wish they knew about the work I do; what I wish for is that children will be introduced to vinyl at an early age because it's not only fun but also very educational. It's physics; it's art; it's culture. You might even say it is a very mindful experience, putting on a record and taking the time to listen to it instead of the hectic time they grow up in. And from a commercial point of view, children might be the record buyers of the future, which will keep our industry alive for many more years to come.

Ren Harcar | Gotta Groove Records
Quality Assurance
Cleveland, OH USA

Ren works in quality assurance at Gotta Groove Records in Cleveland, OH. Gotta Groove operates eight machines and houses the stunning, innovative works of Wax Mage Records. Ren spends her workday listening to and visually inspecting the records they press. Working alongside the press operators, she communicates how to adjust the machines to improve quality and reduce scrap in real time. Ren's goal is to ensure that each pressing sounds and looks as good or better than the initial test approved by the customer. She also recently started quality-checking test pressings.

HOW DID YOU GET INTO YOUR INDUSTRY AND WHAT MOTIVATED YOU TO GET INTO IT?

I found myself at Gotta Groove entirely by chance! I heard about a job opening in Quality Assurance (QA) and applied on a whim. Now five years later, I can't imagine working anywhere else. I've never been a musician, but I am a visual artist and former punk teenager. The world that exists in the overlap between music, art, and design is one that I've gravitated to over the years.

What is your favorite thing about your job and what made you fall in love with what you do?

Gotta Groove is also the home of Wax Mage Records, an all-custom vinyl label founded and operated by our very own Heath Gmucs. Wax Mage is renowned in the industry for really innovative, elaborate designs; what Heath and the rest of our press ops are doing takes vinyl to the next level, and it's truly an art form. Not just the designs themselves but the experimentation and problem-solving that goes into crafting "recipes" for custom Wax Mage builds, ensuring that the finished products still meet our rigorous quality standards. Days that I get to QA Wax Mage customs are my favorite!

WHAT ADVICE DO YOU HAVE FOR SOMEONE WANTING TO PURSUE THIS CAREER?

Every pressing plant is different, but do your research, and don't be afraid to ask questions! Reach out directly to places that interest you. Quality assurance is a niche within an already highly specialized industry, so knowing the right people is important.

> „ Effective communication is crucial across the board. Be assertive and trust your own knowledge and experience, but also understand that making vinyl is a constant learning process, so be receptive to that.

WHAT IS SOMETHING YOU WISH MORE PEOPLE KNEW ABOUT THE WORK YOU DO?

Usually, when I talk about my job, people will joke around like, "I wish I could get paid to listen to music all day!" Listening is obviously a huge part of what I do, but it's a much more intensive sort of listening than how I listen to records at home. There are about a million different issues that can cause records to fail our quality standards: noise anomalies like pops, stitching, nonfill and trimming noise, scratches, blemishes, skipping and wobbling, label imperfections, color purity, center hole distortion, poorly trimmed edges, alignment, flatness – and I'm checking every record that comes off my presses with all of those things in mind.

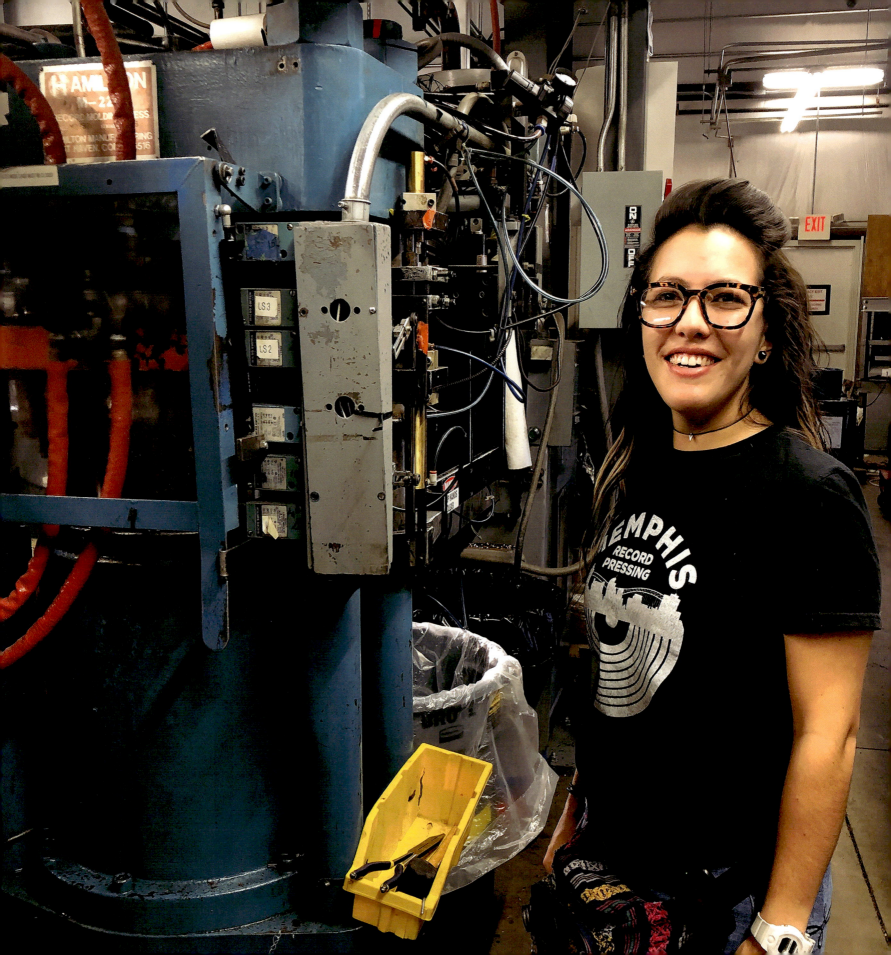

Brianna Orozco | Memphis Record Pressing
Production Supervisor
Bartlett, TN USA

Brianna may have found her way here by chance but has stuck with it. She has worked at Memphis Record Pressing (MRP) since 2015. Brianna started as a press operator and has worked her way up to the 2nd shift Production Supervisor. She oversees all areas of production and ensures that quality products are made in a timely fashion according to their schedule.

How did you get into your industry and what motivated you to get into it?

Pure chance. When I first heard about the company, it was right down the street and very convenient. Once I stepped foot in the doors, I was hooked. I got to see so much fun artwork while assembling and packaging records, and I was curious about the production process and slowly worked my way through all positions that take part in production.

WHAT IS YOUR FAVORITE THING ABOUT YOUR JOB AND WHAT MADE YOU FALL IN LOVE WITH WHAT YOU DO?

Making music, in a sense, by smashing plastic. It's pure sorcery and completely intoxicating. I mostly love the people that wind up here. It's quite a melting pot of all sorts of characters that definitely put the "fun" in "dysfunction." I love my job!

WHAT ADVICE DO YOU HAVE FOR SOMEONE WANTING TO PURSUE THIS CAREER?

> Do it. If you're intrigued, do some research. You'll find out all sorts of cool stuff about the process of making vinyl. If you're close to a record-pressing plant, apply!

The process is mystifying, and I always learn new things every day. Most of my early knowledge came from my old boss, Brian Nickol, who taught me all the things you can't Google an answer for. He's who helped my passion turn into a roaring flame. He was so knowledgeable about *everything* and only wanted to spread that love and knowledge to those who cared to listen. I also simply love music, all varieties, so getting to be a part of someone else's musical enjoyment is quite a pleasure.

WHAT IS SOMETHING YOU WISH MORE PEOPLE KNEW ABOUT THE WORK YOU DO?

How it works! Most people don't even realize what a hot commodity vinyl is nowadays.

Karen Emanuel | Key Production
CEO
London, UK

Karen is the CEO and founder of Key Production Group, which she founded in 1990. Key is a creative agency that specializes in the design and manufacturing of bespoke products for the music and other industries, including, of course, vinyl. She is also the founder of a non-profit organization in the UK called Moving the Needle.

HOW DID YOU GET INTO YOUR INDUSTRY AND WHAT MOTIVATED YOU TO GET INTO IT?

My love for music from an early age, going to gigs and "taping the top 40". At University, I booked the Indie bands across two venues and the DJs for University events; I DJ-d a bit myself. After traveling for a year, I returned to London and decided the music industry was where I wanted to be. I wrote my CV for the first and last time and got a job as the receptionist at Rough Trade Distribution, later I moved into the Production Department, then became Head of Production within a couple of years. I was asked to be made redundant and, from there, set up Key Production with my redundancy. That all usually takes 20 minutes and a PowerPoint to relate!

WHAT IS YOUR FAVORITE THING ABOUT YOUR JOB AND WHAT MADE YOU FALL IN LOVE WITH WHAT YOU DO?

The people, the teamwork and relationships, the music (obviously). Touching the finished product, knowing the work that has gone into it and how much pleasure it gives the recipient. I still feel blessed every day to be part of this vibrant industry.

WHAT ADVICE DO YOU HAVE FOR SOMEONE WANTING TO PURSUE THIS CAREER?

> Follow your passion. Know your numbers. Be authentic. Ask questions. Build your network.

Speak to me! Check out the women's networks and groups for help and advice. I am also the founder of Moving the Needle: www.MTNnow.com

WHAT IS SOMETHING YOU WISH MORE PEOPLE KNEW ABOUT THE WORK YOU DO?

How long have you got?! Making physical products is a complicated process, vinyl especially. So much expertise goes into every detail of a release, from the mastering, design, printing, and pressing. There are so many moving parts. Also, this process involves a long and complicated supply chain that has recently been extremely challenging. At Key, we present "Production 101," which helps people understand all the processes using experts, videos, and "show and tell." The more knowledge people have, the easier it is for them to understand why things can take a long time, where the potential pitfalls are, how to get the best and most sustainable product possible and how to manage expectations.

CHAPTER FOUR

DISTRIBUTION

A distributor in the vinyl industry, when it boils down to it, is just like any other: an entity that buys certain products or product lines and sells them to the customer or consumer. They are the middle (wo)man. Music distribution has a core role in the music business, at the very center of how we purchase our favorite albums. The internet changed music distribution in every possible manner. This discussion will focus on the physical format. But let's take a moment to recognize that illegal music-sharing through Napster and other digital platforms became prevalent in the late 1990s and early 2000s. The practice decreased revenues by about half, forcing the industry to adapt and change in ways never previously imagined. The music business still wrestles with these rippling effects today. Although consumers now pay to download music legally, the digital age changed how music is purchased, listened to, and shared. In 2019 we began to see vinyl records selling more units than they had since 1989, according to the RIAA (Recording Industry Association of America), and resumed their place as the dominant physical format.

Some of the earliest distributors in the world of physical media were Edison, Victor, and later Columbia. These companies formed and led the production and early distribution of recorded music. During this time of rapid growth for the music industry, nationwide networks in the United States between vendors and retail outlets emerged, connecting record stores and labels to these distributors and distribution hubs. Even though it may seem like shellac and vinyl were where distribution began for the music industry, it started even earlier. Before recorded music and radio, sheet music was a hot commodity; and made popular compositions of the time more accessible to those who could read and play an instrument. The first copies appeared around 1473, setting in motion the beginnings of the music industry's distribution chain and allowing for popularity to spread quicker and farther in a tangible way than ever before. By the 20th Century, distributors shifted focus to recorded sound, starting with shellac 78 records. Shortly after, record stores began to open their doors worldwide. Records being the dominant format for listening to music at home and playing on the radio, stores were at their peak from the 1950s through the early 1980s. Record stores drove the importance of distribution companies forward around the world.

Record labels deal with distribution companies, who then sign agreements with retail stores to get records into the hands of you, the consumer. The benefit to the label is that the distributor can provide a more extensive reach to consumers working with retailers both regionally and abroad. The agreements that distributors work on with record labels and sometimes directly with artists allow the distributor the right to sell products at wholesale cost. The distributor negotiates a percentage of the sale from each unit sold and pays the label/artist the remaining balance. A distributor's cut of each album sale drives them to push promotional campaigns and enthusiasm in record stores. Distribution companies' representatives focus on selling and marketing records to all their retail partners, including independent record stores, big box chains, and online stores.

Traditional distributors store or stock warehouses with releases, keep track of invoicing stores, record payments, ship to the retailer, and collect returns. In addition to their primary focus of getting records into stores, distributors often provide additional services. Client and customer support for manufacturing errors and even payment processing according to complex contractual agreements are ways distributors increase their value to both artists and labels. Sometimes they also bridge the gap between the physical and digital worlds to provide digital distribution to streaming services.

In most cases, a record label oversees the creative process, works directly with the manufacturer, and then provides the distribution company with the finished product. There are occasions when distributors offer combination manufacturing and distribution deals. In these agreements, the distributor handles and pays for all upfront manufacturing costs of an album. As sales come in, they keep the revenue until the recuperation of all upfront costs then they begin to pay royalties to the rights owner. These are called Pressing & Distribution (P&D) Agreements.

As distribution companies manage everything from shifting production timelines to marketing campaigns, sales trends, and inventory logistics, it is no surprise that supply chain issues cause massive headaches. With so many pieces to the puzzle, the impact of manufacturing delays can ripple significantly throughout the process and impact the consumer getting their records. The ability to manage these evolving challenges makes distributors valuable, as they leverage their relationships with retailers and the ability to keep the records coming.

Amanda Schutzman | All Media Supply
Sales Representative
New York, NY USA

Amanda is a Sales Representative for All Media Supply, a one-stop distributor that supplies record stores with New Release and Reissue LPs, CDs, and more. She works with over 200 record shops all over America and is proud to call some of them her best friends. Amanda was the manager of a small shop on Long Island, Needle + Groove Records, when they opened in 2018, and still helps out as the part-time buyer there a few days a week, in addition to being their sales rep. Prior to the opening of Needle + Groove, she was the buyer at Generation Records in NYC. She also runs the Vinyl Revolution Record show with her father four times a year in various New York locations.

HOW DID YOU GET INTO YOUR INDUSTRY AND WHAT MOTIVATED YOU TO GET INTO IT?

My father owned a popular record store in Valley Stream, NY, specializing in Rock. More specifically, rare and imported Punk and Metal LPs. It opened in 1982 and closed when I was 18 in 2008. Slipped Disc Records hosted many bands throughout that time, doing in-store signings with artists like Slayer, Metallica, Motörhead, Wasp, Ratt, and many more. I was raised in a record store and more or less bred for music retail. I learned how to write by writing receipts and learned math behind a register. Since Slipped Disc Records closed in 2008, my father has been running Vinyl Revolution Record Shows in Brooklyn, Queens, and Long Island. And besides organizing and running his own shows, he also sets up and sells Slipped Disc stock and merchandise at record shows all over New York and New England. I have always loved being a part of the record show community and getting to know record dealers and collectors from all over the world.

WHAT IS YOUR FAVORITE THING ABOUT YOUR JOB AND WHAT MADE YOU FALL IN LOVE WITH WHAT YOU DO?

My favorite thing about my job is getting to know the customers. I've always thrived in customer service jobs, and selling something you really love, especially when you're a consumer yourself, makes it so much fun. Now that I'm a Sales Representative for a large distributor, I'm on the retail side less than I used to be. My customers are the stores themselves. But because I spend so much time shopping, traveling to record stores, and still keep in contact with all my regulars from the shops I used to work for, I live the best of both worlds. It's easy for me to find you the records you want, whether you're a store or a personal collector. And no matter who you are, I'm happy to do it.

WHAT ADVICE DO YOU HAVE FOR SOMEONE WANTING TO PURSUE THIS CAREER?

There are a lot of ways to pursue a career in vinyl. But over the years, I've learned it's just about getting to know people.

> **"** Get to know the people working at your local record stores, dealers at record shows in your area, and other collectors. Sooner or later, you'll hear about a position that opened in something.

Whether it's working a day a week at a small store, a management position at a big store, sales, purchasing, selling online, at shows, or for a distributor or label. The opportunities are out there. I realized people just fall into positions over time. Just put the work in, and more work always seems to find you. Many people have inspired me to continue doing what I'm doing over the years. And the vinyl community has built a strong support system to help you pursue anything you really set your mind to.

WHAT IS SOMETHING YOU WISH MORE PEOPLE KNEW ABOUT THE WORK YOU DO?

I know the thought of working in a record store sounds like a lot of fun … and it is! But it's also constant hard work, and you're always learning something new each day. It's just as mentally demanding as it is physically demanding, and you really need to apply yourself to excel at both. I work almost seven days a week. Monday through Friday, I work my usual 9 to 5 as a Sales Rep and then head over to the record store and price the new orders on weeknights. On the weekends, I'm working or running a record show, updating store inventory, taking special orders, or working on a store website. There is always work to be done and new things to learn. But that's also what makes me love it.

Jocelynn Pryor | AMPED Distribution
VP of Marketing
Long Beach, CA USA

Jocelynn is the Vice President of Marketing at AMPED Distribution. AMPED™ provides solutions for physical and digital music distribution across the globe. Jocelynn and her team help record labels and artists to strategize their go-to-market efforts and quarterback the retail setup and DSP pitches.

HOW DID YOU GET INTO YOUR INDUSTRY AND WHAT MOTIVATED YOU TO GET INTO IT?

I worked in a mall shoe store in high school and quickly learned that the store across from me was WAY cooler than the one I was in ... so I applied. And so it began ... I started in record retail in '91 at a Sam Goody mall store in Downey, CA. My friend's mom (Debbie English) was a designer of the retail stores, and I heard in passing that she went to the corporate office and talked to buyers and marketers. From then on, I realized this was a big business, and if I could get to a corporate office, I could have a career in music. I worked my way through record retail at Borders Books & Music and Wherehouse Music, where I got to team up with the famed Violet Brown and Transworld (f.y.e.) before making the leap to wholesale/distribution/eCommerce retail at Super D. Super D eventually bought Alliance Entertainment. We merged the two indie distribution arms of the company together and called it AMPED Distribution. I would be remiss if I didn't add that I grew up in a very musical family. My mom is a pianist, my dad led music ministry at church, and my sister and I sang in choirs growing up. We even formed a family quartet at one point and toured churches singing hymns in four-part harmony! It was the literal "Daddy sang bass, momma sang tenor, me and little sister...." All that to say, music was in my blood.

WHAT IS YOUR FAVORITE THING ABOUT YOUR JOB AND WHAT MADE YOU FALL IN LOVE WITH WHAT YOU DO?

I love connecting people, and marketing is just that. It's connecting would-be fans with the music/artist. There is nothing more satisfying than introducing someone to music that eventually becomes the soundtrack to their lives. It's a BIG industry but a very small world, and I love keeping track of all my colleagues, past and present. It is a family of sorts, and there are a lot of real characters in the family! I came up at a time when fewer women were in positions of power. I also came up at a time when the last bit of the old school old guard "break-your-knees" kind of industry was fading out. The first thing I love about my job is seeing the fruits of my labor. There are so many artists today whose careers I had a hand in: whether it was playing the name that tune game at a retail store in the mall as a cassette-single-specialist, or playing videos of new music to a bunch of retailers at a convention, or pressing labels for pitch decks for DSPs. Like a mama bear, my heart goes pitter-pat when an artist whose project I have worked charts or they go on to become a megastar.

WHAT ADVICE DO YOU HAVE FOR SOMEONE WANTING TO PURSUE THIS CAREER?

> " Listen and sponge-up information. Remember that the only thing you have is your name, so make sure you always keep it on the level because people remember.

Use the amenities meaning the A2IM or Grammy U mentorship programs (I wish we had that sort of thing when I was coming up). Keep up with people you know. Be as well-rounded as you can be. If you are a publicist, know HOW what you do affects the supply chain. If you are a manufacturing plant, know HOW what you do affects the touring industry. If you are an artist, know how your decisions affect every indie retailer out there, etc. We are all connected, and an artist must get it right on all sides of the business to have a sustainable, pay-the-mortgage career. We ARE saving lives, one ear at a time! And for goodness sake, have FUN.

WHAT IS SOMETHING YOU WISH MORE PEOPLE KNEW ABOUT THE WORK YOU DO?

I love public speaking and being a panelist because I can talk about the music business all day long! I spend a fair amount of time mentoring young women in the business, so they have a leg up that I didn't have coming up. I'm also a fan of advocacy work; just last week, I spoke to my congressman's office about the HITS ACT and the RAP ACT.

Christie Coyle | Redeye Worldwide
Senior Account Representative
Seattle, WA USA

Christie is a Senior Account Representative with Redeye, an independent physical and digital distributor representing over 250 label partners globally. Based in Seattle, she works closely with independent record stores, online retailers, one-stop partners, and other non-traditional accounts throughout western North America. Christie is involved in many facets of Redeye's business, from ensuring accounts are covered on new releases and scheduling daily shipments with the warehouse to coordinating marketing opportunities with record stores, such as merchandising or in-store performances.

HOW DID YOU GET INTO YOUR INDUSTRY AND WHAT MOTIVATED YOU TO GET INTO IT?

Music has always been near and dear to me, but having a radio show at WUAG in Greensboro, NC was definitely the jumping-off point. At the station, I was introduced to a treasure trove of music that I may have never otherwise experienced. Playing bands (as well as seeing them perform live) like Trans Am, Battles, Beach House, Toro Y Moi, and more made me realize that I wanted to be around music in some capacity. Redeye, which is based in North Carolina, was introduced to me by a close friend. I applied and interviewed for various positions over the course of 4 years, from entry-level to management, and finally landed the account rep position in 2013. I relocated to Seattle in 2015 to represent our company on the west coast, which allowed me to better serve our accounts in the western time zone.

WHAT IS YOUR FAVORITE THING ABOUT YOUR JOB AND WHAT MADE YOU FALL IN LOVE WITH WHAT YOU DO?

Without a doubt, my favorite part of the job are the bonds that I have formed with customers over the years. Many of the people whom I work with on a daily basis have become close friends. That level of camaraderie makes even the most frustrating days easier because I know they'll be there to show appreciation and support, make light of the hard times, or simply share in our love of music.

WHAT ADVICE DO YOU HAVE FOR SOMEONE WANTING TO PURSUE THIS CAREER?

Be persistent! With Redeye, I applied for every position available, from social media director to warehouse worker to marketing manager.

> "As with any profession, networking is key. Even having one contact can help get your foot in the door.

Both the Women in Vinyl and A2IM job boards are also fantastic resources for searching for tons of available positions throughout the music industry.

WHAT IS SOMETHING YOU WISH MORE PEOPLE KNEW ABOUT THE WORK YOU DO?

A distributor has the unique role of acting as a bridge between our label partners and the independent retail community. Working with so many different genres and artist profiles allows Redeye to inform our friends in indie retail of music they may not otherwise be aware of. It's about educating buyers who have so much on their plates as it is that some often don't have the time to dive deep into our extensive catalog. The core function of my role is to highlight and champion the releases that will have an impact at their store and within the local community.

Shelly Westerhausen Worcel | Secretly Distribution

Head of North American Sales & Marketing

Bloomington, IN USA

Shelly is the Head of North American Physical Marketing and Sales at Secretly Distribution. On a macro level, she manages a team of marketing and sales reps that handle selling and marketing all physical products to hundreds of US and Canadian outlets, including indie record stores such as End of an Ear in Austin and Seasick Records in Birmingham to big box chains like Barnes and Noble, Target, etc., to online retailers such as Turntable Lab, and finally subscription services like Vinyl Me, Please.

HOW DID YOU GET INTO YOUR INDUSTRY AND WHAT MOTIVATED YOU TO GET INTO IT?

I've always had a fierce passion for music. When I was young, I took vocal, guitar, piano, saxophone, and clarinet lessons but also struggled with really bad stage fright. In high school, I had a very frank conversation with one of my lesson instructors, where he basically said it was obvious that I have a passion for music, but I didn't necessarily have the talent it would take to pursue music professionally. He suggested I check out the business side of the industry. I was intrigued, and this conversation led to me flying out to LA at 15 to attend a two-week-long Music Business camp, and I fell in love. I hopped around from Belmont University in Nashville, TN, and Indiana University in Bloomington, IN, to get a degree in Art Administration with minors in music and music business. During my college years, I interned in several different areas of the business – with a management company, a booking business, and the warehouse at this distribution company. I fell in love with the indie side of the industry and secured a job with Secretly before even finishing college. Since then, I've worked my way up from intern to shipping in the warehouse. Then on to assist the sales team, and eventually, I became a sales team member. And then, I moved on to a marketing team and finally landed here managing our sales and marketing team. We currently work on hundreds of release campaigns every year from over 100 independent labels, management companies, and artists. When I'm not working on music, I spend my free time making vegetarian meals in the kitchen and wandering antique stores.

WHAT IS YOUR FAVORITE THING ABOUT YOUR JOB AND WHAT MADE YOU FALL IN LOVE WITH WHAT YOU DO?

I love working with people who are as passionate about music as I am every day. It's also super rewarding to know you've directly impacted the success of a campaign for an artist. Knowing that the hard work you put into a campaign directly impacted how well an album performed and will help move an artist's career forward is always such a positive feeling.

WHAT ADVICE DO YOU HAVE FOR SOMEONE WANTING TO PURSUE THIS CAREER?

> " Stay curious and try to make connections along the way.

As an introvert, I always hated hearing the "network more" advice, but there are ways to meet people and explore the industry without being too forward. If you can, take programs in school that will offer resources for finding industry jobs. Try out a few entry-level positions or industry internships to see what resonates with you. Go to shows and your local record stores to become a part of your community's music scene.

WHAT IS SOMETHING YOU WISH MORE PEOPLE KNEW ABOUT THE WORK YOU DO?

Distribution is definitely a part of the industry that can be easily overlooked – perhaps because people don't always understand what we do. Our job is so fun because we spend all day talking about music to a wide range of industry folks. Depending on the position within distribution, we work with record stores, labels, artists, managers, digital service providers, and manufacturers. Who wouldn't want to spend a large part of their day talking about and getting people excited about music?

CHAPTER FIVE

RECORD LABELS

The term "record label" was coined from the center label of a vinyl record, which prominently displayed the name of the company representing the artist. A record label is a company that manages, coordinates production, manufactures, distributes, markets, promotes, and handles copyright for sound recordings of musical talent. They sometimes accept submissions or otherwise scout and develop new artists. The term you may know of as A&R, or "artists and repertoire," is an abbreviation for that part of the process.

Today's music industry allows artists to reach audiences directly through social media. However, recording artists have traditionally relied on their labels to broaden their consumer base, market their albums, and promote singles on the radio. While not all artists have a record label, when they do, the label is critical in getting their albums manufactured and into the hands of fans through distribution companies, as mentioned in Chapter 4.

You may notice certain companies repeatedly at the forefront of history by this point in the book. This chapter is no different; Edison Records was the first "record label," but they made wax cylinders. The next record label has a name you will likely recognize, Columbia Records; they were at the forefront of popularizing the record and are still a prominent label today.

When you think of iconic record labels, you might think of the iconic image of a dog with his ear up to the cone of a **gramophone** listening to his master's voice. RCA Victor used the "his master's voice" image to popularize the gramophone while touting the clarity of the sound of their records. Victor was well known for higher quality, better-sounding records than Columbia. Both labels began by releasing Opera recordings, some of the first records sold.

With vinyl as the main format, the music industry continued to grow; the 1960s brought the rise of CBS and Warner Brothers. The 1970s saw prominent labels like Capitol Records, Polygram, and EMI emerge. By the 1980s, vinyl was at its peak, and the significant labels included the "big six." It was a term that referred to: EMI Sony Music (formerly CBS Records), Warner Music Group, BMG (formerly RCA/Ariola International), Universal Music Group (formerly MCA Music), and PolyGram. Over the years, these labels have continued to merge, leaving what we now consider the three major labels: Warner Music Group, Universal Music Group, and Sony, which contain sub-labels that they control. A major label is any label with greater than 5% of the global market share.

Whenever you have large entities like this, it calls for a balance of power, which is what happened with the rise of the independent record label. An independent label is not funded by or affiliated with the major labels and handles its own marketing, distribution, and funding. Indie labels began by promoting smaller genre subsets or groups outside of the mainstream, like Sun Records in the 1950s or Motown in the 1960s to Sub Pop in the 1990s, which had huge impacts on defining certain eras within music history. Though a smaller reach, their impact rippled through local or genre-specific communities. They often predicted trends in politics, fashion, and beyond, including trends the major labels and mainstream artists would capitalize on later. Independent labels are crucial to serving fans of particular subgenres and afford these musicians an opportunity in the business.

The independent record label's part in producing an album release requires a lot of collaboration with an artist. They work out what size of a run the physical release should start with and which projects are suited to specific formats. Regardless if the label is an indie or a major, they carefully consider the budget, art, design, how many records the artist will sell, and how to best represent and promote the artist and each project. Part of this means looking at the artist's fan base and how they are consuming music. Will the artist be able to sell out of their vinyl release? If yes, the record label will look for a pressing plant or **vinyl broker** to work with to create the artist's vision.

Sometimes the artist has a specific vision for the album artwork and packaging. Other times, the label takes the lead in coordinating and hiring photographers, illustrators, and the like to produce visuals and see which pressing plant is best to execute that vision within budget. Labels decide whether to use a particular pressing plant or go through a vinyl broker. A vinyl broker works as a bridge between the label and the pressing plant to handle many logistics that may seem overwhelming for an artist or small label.

After they place the order, the label will work with distributors and create a marketing plan to set a release date. The **release date** is the anticipated date that the record is to be released, and the **street date** is when the physical product will be available. Nowadays, release dates are sometimes staggered, with a digital release first and physical media to follow. The label may also set up and use tour dates to promote the release. Indie labels are often more artist-friendly and typically offer more considerable artist royalties. They also tend to be owned by other artists, and they tend to take a personal interest in the quality of the releases and share a personal connection to their artists' work. Whether the label is major or indie, it can play a prominent role in supporting an artist's vision and exposure.

Désirée Hanssen | Lay Bare Recordings
Owner
Nijmegen, The Netherlands

Since 2013, Désirée has been at the helm of Lay Bare Recordings, an independent and boutique-styled record label that supports bands that she believes add value to the unconventional and underground music scene. She sees Lay Bare Recordings as a tool for bands to reveal their music so that as many music lovers as possible can discover it. She is currently working on a new concept, Sound of Niche (SoN), a collaboration between Lay Bare Recordings and Galloway Studio, focusing on the power of community, collaboration, and challenging creative extravaganzas. She has also joined forces with Alan Pride to represent Lay Bare Recordings in the UK, working closely to develop and enhance her plans for the label's future.

HOW DID YOU GET INTO YOUR INDUSTRY AND WHAT MOTIVATED YOU TO GET INTO IT?

In the nineties, I worked in a venue called "Doornroosje" (www.doornroosje.nl), where I met many friends and music lovers. Almost everyone from that time is still related somehow to the independent and more unconventional music scene. In this venue, I saw so many superb and legendary bands. And it was also at this place I became friends with many die-hard live fans and people who got into the music business. We still have one thing in common, our passion for music, especially for vinyl and gigs. If people want to know more about my history related to music, check out this awesome book: "Passion for Vinyl" by Robert Haagsma (www.passionforvinyl.com).

I am an avid record collector, love to attend concerts, and always look for a new take on music experience. At home, I share my passion for music with my partner, Manuel Tinnemans; he is the creative mind behind the dark art from www.comaworx.com. There is always a record spinning in one of the rooms.

Besides running my record label, I have a day job as a counselor and trainer. I coach and motivate people who have lost a lot in their life. I empower them to see their talents and show them how to use their skills to get a better feeling and grip on their life again. I learned a lot from working with juvenile delinquents who already had little perspective at a young age. I mentored them to get back on track; one big part of this was finding their talents and creativity.

WHAT IS YOUR FAVORITE THING ABOUT YOUR JOB AND WHAT MADE YOU FALL IN LOVE WITH WHAT YOU DO?

Seeing the most awesome music and watching the bands work hard to get their music out there, it was about time to contribute to those musicians. So I wanted to work with them and make it possible to get their songs on vinyl on a broader scale. Like the name of my label: showing the world those precious things that have been "underground" too long. Those exceptional gems that are worth exposing their brilliance to the surface. "To LAY BARE something."

WHAT ADVICE DO YOU HAVE FOR SOMEONE WANTING TO PURSUE THIS CAREER?

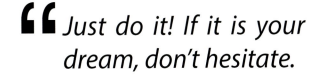

> **" *Just do it! If it is your dream, don't hesitate.*

Women like Deborah Harry, Frida Kahlo, Betty Page, Billie Holiday, and Dolly Parton encouraged me. All inspiring human beings who didn't limit themselves just because people won't accept that they can or could do something.

But be prepared to invest some of your savings. DIY isn't about money-making, and I will be very happy when the day comes that I can break even. So like I said, just do it! It brings you friendship, music, conversation, joy, and some great results: shiny pieces of great wax!

WHAT IS SOMETHING YOU WISH MORE PEOPLE KNEW ABOUT THE WORK YOU DO?

Working as a label owner, you are acting in many different roles. Sometimes as a motivator, mediator, negotiator, creative consultant, PR, booker, manager, and influencer, but above all, a companion that collaborates with great artists to realize that their music reaches out to the public. The advice I can give is to accept that it becomes a big part of your lifestyle. Especially when you got bands overseas, working and contact hours are different than your daily life planning—irregular schedules. A new release is never the same. And stick to an open mind, like "think outside, no box required."

Katy Clove | Merge Records
Production Manager
Durham, NC USA

Katy is the Production Manager at the independent record label, Merge Records. As the production manager, a typical day might include scheduling mastering for an artist; ordering lacquers; checking masters; checking test pressings; inspecting finished goods; communicating with artists, print vendors, and vinyl pressing plants; pre-flighting files; checking proofs; requesting estimates; budgeting; brainstorming for packaging and promotional materials; creating purchase orders; placing and tracking orders; reconciling invoices. For Katy, the production schedule begins with the artist delivering mixes for mastering and ends with the receipt of finished goods.

HOW DID YOU GET INTO YOUR INDUSTRY AND WHAT MOTIVATED YOU TO GET INTO IT?

I have a fine arts degree in photography. In 2001, after moving to North Carolina from Oregon, I got an entry-level job in the design and production department at a scholarly book publisher. That job introduced me to print production, bookmaking, typography, graphic design, proofing, and color management. Every opportunity I've had since can be traced back to the people I met while working there. The "logical left-brain, creative right-brain" myth never really resonated with me; even while pursuing an art degree, my love of art, math, and science didn't feel like competing interests. I found that I was just as interested in the methodical and repetitive side of production as I was in the creative and intuitive practice of making art.

Shortly after leaving that job, I was invited to come back and sit in for one of the designers while she was on maternity leave. After that immersion, I began to freelance as a designer—mostly book design and collateral—and, because I had so many musician friends, music packaging. Somehow, despite never having been a musician, music has been integral to every phase of my life since I was a teen.

In 2007, via word-of-mouth and mutual friends, I was again recruited to fill in for a designer on maternity leave—this time, it was for Merge Records. The timing coincided with Merge beginning to work on the print for their 20th anniversary. I spent most of my time working on a box set that included 17 CDs, a poster, and a book. I continued to freelance for Merge throughout the life of that project. In 2009, in need of a full-time job, I began working at an ad agency doing prepress, color management, quality control, and software support. I managed the large-format printers and set up the in-house proofing system. I worked there for seven years before moving into my current position at Merge Records.

WHAT IS YOUR FAVORITE THING ABOUT YOUR JOB AND WHAT MADE YOU FALL IN LOVE WITH WHAT YOU DO?

My favorite thing about my job is the people. The artists, Merge staff, and our vendors (some of whom we have worked with for over 30 years) inspire and motivate me every day.

WHAT ADVICE DO YOU HAVE FOR SOMEONE WANTING TO PURSUE THIS CAREER?

You will be rewarded with generosity and opportunity if you are humble, gracious, curious, and kind. Be resourceful and do your research.

> **❝ Ask questions and educate yourself. Develop a versatile skill set and stay current. Listen.**

WHAT IS SOMETHING YOU WISH MORE PEOPLE KNEW ABOUT THE WORK YOU DO?

Managing the expectations of artists, colleagues, and vinyl consumers is extremely challenging. I wish more people understood the labor and craft that goes into making records beyond the recording, mixing, and mastering stages. That knowledge can enrich the listening experience and encourage more forgiving, empathetic consumers.

Riley Manion | Secretly Group
Production Director
Bloomington, IN USA

Riley was born and raised in Bloomington, Indiana, the hometown of Secretly Group. She is the Production Director for all associated labels – Dead Oceans, Jagjaguwar, Secretly Canadian, Saddest Factory Records, The Numero Group, Ghostly International, and drink sum wtr. Riley oversees the production of all physical media, the majority of which is vinyl.

HOW DID YOU GET INTO YOUR INDUSTRY AND WHAT MOTIVATED YOU TO GET INTO IT?

Music has been a big part of my life since birth. My dad was in community radio and had my sister and me on the mic as toddlers. I DJ'd throughout high school and my 20s, interned at a record label in high school, collected records, went to as many shows as I could, visited record stores in every town I traveled to, and was surrounded by people running record labels, including the Secretly folks. I knew I wanted to be involved in the music industry, but I wasn't sure exactly how. I focused on photography for a while, but it wasn't quite right. When a position with the physical media team at Secretly opened up, it felt like the perfect fit.

WHAT IS YOUR FAVORITE THING ABOUT YOUR JOB AND WHAT MADE YOU FALL IN LOVE WITH WHAT YOU DO?

Seeing the physical copies land is always a joy. Knowing how much work went into getting the details right and achieving the deadline makes it even more satisfying. I love seeing abstract digital sketches or verbal conversations morph into a beautiful object. Working on Numero's Cabinet of Curiosities series has been a highlight. They would say something like, "we want to make a jacket that expands into a pyramid," or "this needs to actually smell like a pack of menthols," and then put their full trust in me to find the right partners to bring it into the world! In the end, all of this work revolves around music. I really fell in love with the role and this corner of it, specifically because of the amazing artists on the labels.

WHAT ADVICE DO YOU HAVE FOR SOMEONE WANTING TO PURSUE THIS CAREER?

Looking back, my inspiration came from my community. The Bloomington scene in the late 90s and early aughts was inclusive and encouraging for women. We felt safe at shows, we enthusiastically made zines, and although I never played an instrument, there were many women and LGBTQ+ friends in bands. It was a special era and gave me the space and opportunity to experiment with what I was interested in.

> **"** My advice is to get involved early and often. Try different things. It's hard to predict where the spark will come from.

Volunteer at the radio station, sign up for Girls Rock, organize a show, interview a band, intern at a label, write record reviews, and record sounds that interest you. There are so many different roles in the music industry. Trying out many types of work will help you find your passion and meet people. For my role specifically, I'd encourage people to collect physical media. Study the packaging and get familiar with the details. Production is full of small details to keep track of.

WHAT IS SOMETHING YOU WISH MORE PEOPLE KNEW ABOUT THE WORK YOU DO?

Getting a record in the world takes a tremendous amount of team effort and time. Growing up, I was always blown away by the length of film credits, but I noticed that album credits focused mostly on performance and recording. Of course, I had no idea how many people were involved, but now that I'm in the industry, I see we could fill pages of every release's liner notes with the people involved. Everyone from the project team planning with the artist(s), to the art team realizing the vision of the band, to my team turning that vision into a physical product with our manufacturing partners, to the distribution team getting the album into the world, to the marketing team promoting it, and finally to all the record stores stocking it for the fans by release date (plus many other steps in between!). It's a wild chain, and I love working hard to transform it into a physical product that people can hold in their hands.

Katrina Frye | Lauretta Records
Founder & CEO
Los Angeles, CA USA

Katrina Frye has embarked on a new approach to "old traditions" in the arts economy. Launching Mischief Managed in 2013, she experimented with a subscription-style company offering management and marketing support for artists of all mediums. After working with over a hundred artists through workshops, lectures, and one-on-one coaching, Katrina wanted to narrow her focus within the music industry. In 2020 Katrina founded the indie music label Lauretta Records from her living room floor. With a decade of experience behind her, Katrina wanted to sign like-minded artists who were ready to work but had not found an open door yet. She has lectured at institutions like CBU Riverside, CSU Long Beach, CalArts, PCC, and more. Katrina has brought her years of experience to the next generation of artists and creatives.

HOW DID YOU GET INTO YOUR INDUSTRY AND WHAT MOTIVATED YOU TO GET INTO IT?

I was studying visual arts and planning lots of parties and realized art and commerce go hand in hand. Then the guy I was dating started touring his band and needed help with booking, invoicing, payments, and merch, so I got a front-row seat to the entire music industry.

I was seeking to create space for underestimated artists and musicians and was determined to engage indie artists in a sustainable way. Completing the A2IM's Black Independent Music Accelerator Program in 2022, I feel more energized and ready to support the next chapter of music makers. By giving artists education and access to information, I believe empowered artists strengthen our entire industry. With the current six-artist roster, they have placed over 100 songs in tv/film. I have also enjoyed leading conversations and workshops for the United States of Women, Netflix, ArtCenter, LA County Arts Commission, and the Center for Cultural Innovation for the larger entertainment industry. No matter the platform, I am always thrilled to share my message of hope and possibility with emerging artists and creatives around the world.

WHAT IS YOUR FAVORITE THING ABOUT YOUR JOB AND WHAT MADE YOU FALL IN LOVE WITH WHAT YOU DO?

My top three favorite things about being the CEO of an indie music label are that every day is different, each artist has different needs, and the industry is always evolving. I love that you always have to learn, grow, change and evolve with the artist, the industry, and the music.

WHAT ADVICE DO YOU HAVE FOR SOMEONE WANTING TO PURSUE THIS CAREER?

My advice would be don't wait for someone to give you permission to enter this industry.

> **Make your own way and try everything. Don't be too proud to work for free, work on commission, work for trade.**

As you grow in experience, keep evaluating your worth and contribution. Don't be afraid to keep putting yourself out there.

WHAT IS SOMETHING YOU WISH MORE PEOPLE KNEW ABOUT THE WORK YOU DO?

It takes so much time. People want instant results, instant money, and instant success, but those measurements take time.

Julia Wilson | Rice Is Nice
Owner
Victoria, Australia

Julia is the owner of Rice Is Nice, an independent Australian record label which aims to promote unique Australian artists. Julia formed the label in 2008 and has since signed and released the works by a number of artists, the musical genres of which vary greatly. She also started Brain Drain, which is a label management and PR firm, working with select publicity clients and labels managing a few independent record labels. Julia also works as a music supervisor at Nice Rights and Midnight Choir.

HOW DID YOU GET INTO YOUR INDUSTRY AND WHAT MOTIVATED YOU TO GET INTO IT?

I have always wanted to run my own record label. I worked as a music photographer and in a few independent record stores.

WHAT IS YOUR FAVORITE THING ABOUT YOUR JOB AND WHAT MADE YOU FALL IN LOVE WITH WHAT YOU DO?

I enjoy supporting artists; that's always been my drive. I love advocating for people with creative, innovative ideas and wishes and trying to make them practical possibilities. I am very thankful that I work in a space where we have the freedom and support to do whatever we want.

WHAT ADVICE DO YOU HAVE FOR SOMEONE WANTING TO PURSUE THIS CAREER?

> " Talk to everyone, meet with everyone. Work out where you want to be, who you would like to work with/for, and go for broke!

It's tough to be that social, but you will learn from the good humans doing the hard work. You will quickly work out what you do and do not like.

WHAT IS SOMETHING YOU WISH MORE PEOPLE KNEW ABOUT THE WORK YOU DO?

Our job is constantly educating that music, the creatives, and the industry have value.

CHAPTER SIX

RECORD STORES

If you're reading this book, you've likely had the conversation more than once. The one where you tell someone you own, collect, or work in vinyl records, and their response is, "…They still make those?!" While only some people are up to date on what has been considered the vinyl revival, we believe every city or town should have a record store, and even as far back as the 1800s, someone agreed. That wild intrepid soul was Henry Spiller of Spillers Records, considered the oldest independent record store in the world, located in Cardiff, Wales, and founded in 1894. How's that for old-school street cred?

Thomas Edison's **phonograph** received its patent in 1877, and Emile Berliner's gramophone followed in about ten years. You can imagine Mr. Spiller's enthusiasm to open a record store only 17 years later in a small town would seem like an unusual risk to anyone. Luckily this new technology wasn't just a fad. Over a hundred years later, there are still generous, intrepid souls following in Henry's footsteps. They pour their life savings into the dream job of steward and local musical tour guide at record stores across the globe.

It would be another decade before records would be widely available commercially. Early distributors (as mentioned in Chapter 4) consisted of familiar names like Edison, Victor, and later Columbia. In the early days, distributors sold shellac 78s. As we know them today, vinyl records are a very different product. RCA Victor launched the first commercial long-playing record (LP) in 1930. Shortly after, record stores began to open their doors worldwide, reaching their peak in the 1950s through the early 1980s. Records were the dominant format for listening to music at home and playing on the radio.

You may, at this point, be thinking, "Isn't vinyl at its peak now?!" Honestly, it hasn't always been sunshine and endless crate-digging down at your local record store. There were dark days for both record stores and the format during the digital revolution. For a while, there was boundless excitement from the music-buying public for the new technology of **compact discs (CDs)**. Developed by Phillips in 1974, the CD started reaching popularity in the 1980s, significantly hurting the vinyl business from the late 1980s until the early/mid-2000s. Many people within the vinyl industry and the vinyl community credit DJs and small labels for keeping the format alive during this time.

Then, something changed at a meeting in Baltimore, MD; a group of diehard record store lovers started a revolution. Inspired by a Free Comic Book Day event, these indie record store stalwarts put their heads together and created an event to celebrate the humble yet vitally important independent record store, many of whom had never stopped selling vinyl. The event became known worldwide as *Record Store Day* and expanded quickly to thousands of stores and, when needed, multiple days a year. At its heart, it's a party. Each store curates its experience for the day with celebrations touting in-store artist performances, meet and greets, and fundraisers for community non-profits. And, of course, special vinyl releases and promotional offerings you can only get at the record stores. Since the first official Record Store Day event in 2008, vinyl has continued to soar in popularity, inspiring new record stores to open almost daily.

Today with that popularity, anyone can buy and sell records, and not everyone needs a brick-and-mortar storefront to live the dream. Social media and online community sites like Discogs have changed our view of the indie record store and how we purchase vinyl records. While to a lot of us, there is nothing like the experience of visiting a brick-and-mortar store, getting recommendations from the person behind the counter, and flipping through the new arrivals bin, online accessibility opens up a whole new world of relationships and records never before available. We still recommend getting to know your local record dealer and cultivating that relationship; even if you connect through a virtual platform, there is no substitute for great recommendations.

Whether you have a physical location or not, owning a record store is not sitting around all day listening to music, though that is a perk. The ins and outs of the business involve having your finger on the pulse of your community, your client base, and what is popular, just like any other retail environment. Acting as your customer's music guide, making recommendations, special ordering, and stocking what they love is a massive part of the business. That means heavy boxes of inventory, setting up special events, and always staying ahead of the curve. Record store owners decide what to stock in their new and used sections for their local masses. New records are typically ordered from a distributor (as discussed in Chapter 4) and priced as such so that they can make a profit. However, used records can come from anywhere, such as someone selling their collection to the store, estate sales, flea markets, etc. When used records go into a record store, pricing is more subjective, and the owner needs to assess each record, clean it, and grade it appropriately based on the condition to sell it and make a profit. There are a couple of widely accepted grading systems, such as Goldmine and Discogs, with detailed ratings from Mint to Poor and definitions of everything in between.

Make sure to support your local record store and the people behind the counter, to bring you the music you love. Talk with and learn from them.

Lolo Reskin | Sweat Records
Owner & Music Buyer
Miami, FL USA

Lolo is a lifelong Miamian and music lover. She started Sweat Records in 2005 to cater to Miami's huge audience of music lovers, as well as visitors to their tropical metropolis. They carry Miami's widest selection of new and reissued vinyl, as well as used LPs, turntables, accessories, enamel pins, local merch, zines, and more. Sweat Records is a local institution as well known for its regular schedule of all ages in-store events as it is for being a world-renowned record store.

HOW DID YOU GET INTO YOUR INDUSTRY AND WHAT MOTIVATED YOU TO GET INTO IT?

My dad's side of the family are all classical musicians, and thankfully I also grew up next to a legendary Miami roller rink where I fell in love with pop music and DJ culture. I started working at a Virgin Megastore at 16, then left at 22 to open Sweat Records. I was street-repping for labels, DJing, booking bands, and throwing indie club nights. I am grateful to have known very early on that I'd be miserable in any career that didn't involve music.

WHAT IS YOUR FAVORITE THING ABOUT YOUR JOB AND WHAT MADE YOU FALL IN LOVE WITH WHAT YOU DO?

> **"** I love people, and music appreciation/fandom is a universal language that instantly creates connections. We are so happy to be in Miami, where we get customers from all over the globe literally every day.

It is immensely satisfying to have conversations and make new friends over these shared bonds, and 23+ years in retail later, it's still fun to turn people on to new sounds they'll swoon over.

WHAT ADVICE DO YOU HAVE FOR SOMEONE WANTING TO PURSUE THIS CAREER?

I'm deeply inspired by Mitch Kaplan, who founded Miami's iconic indie bookstores Books & Books and the world-renowned Miami Book Fair. He created a community space centered around the love of books which is what we've always strived to do with Sweat and vinyl and the events we do.

WHAT IS SOMETHING YOU WISH MORE PEOPLE KNEW ABOUT THE WORK YOU DO?

The music industry is one of the hardest out there, and we're putting in the sweat (!) because what we do is important. I always say that record stores are the boots on the ground of the music industry as we're putting records in people's hands and spending countless hours poring over new release listings to select the best ones. The experience of coming to a store like ours is irreplaceable, and our staff's recommendations kick any algorithm's ass.

Brittany Benton | Brittany's Record Shop
Owner
Cleveland, OH USA

Brittany is the owner of Brittany's Record Shop; an independent record store specializing in hip-hop, reggae, and soul out of Cleveland, Ohio. Brittany is also a DJ and producer as DJ Red-I and a high school teacher.

HOW DID YOU GET INTO YOUR INDUSTRY AND WHAT MOTIVATED YOU TO GET INTO IT?

My first experience with vinyl was accessing my grandmother's and great-aunts' music collections. Everything from Mary J. Blige, Santana, Mahavishnu Orchestra, Dorothy Ashby, and whatever else I could get my hands on.

I got into the vinyl industry through years of adjacency as a DJ, beat-maker, and general music collector. I was always looking for funky breaks or ear-wormy samples by digging in the crates. A lot of my friends were into wax too. In 2015, I had the opportunity to buy the contents of a record shop facing closure. Less than a year later, I split with my business partner and reopened the shop as "Brittany's Record Shop."

WHAT IS YOUR FAVORITE THING ABOUT YOUR JOB AND WHAT MADE YOU FALL IN LOVE WITH WHAT YOU DO?

I love that music has made a way for me. Music has been a part of my life and how I've made a living for over a decade; as a DJ, producer, and educator. But as a record shop owner, I'm most proud of knowing that people often come to me to connect them with music that they didn't yet know they loved. I love connecting customers with something new and the shared nostalgia of reconnecting them with something that takes them back to a special time in their lives.

WHAT ADVICE DO YOU HAVE FOR SOMEONE WANTING TO PURSUE THIS CAREER?

❝ Don't try to please everyone. Find a niche and serve it.

By doing this, you are building a community that you can always serve because you understand how it is overlooked and, in turn, needs to be cultivated.

WHAT IS SOMETHING YOU WISH MORE PEOPLE KNEW ABOUT THE WORK YOU DO?

It's always changing as I grow. Being a one-woman show often reflects how things are going in other aspects of my life. When I was in my 20s, my work focused on DJing at bigger and bigger clubs and going on tour with my group, but now I am teaching, producing music, stocking the shop, and other things. My work is my life, and I enjoy experiencing it as it evolves in real time. That said, I'm sure that music will remain at the core of it all.

Sharon Seet | The Analog Vault
Curator & Co-Founder
Singapore

Sharon is the curator and co-founder of The Analog Vault (TAV), a record store based out of Singapore. Since its founding in 2015, TAV has sealed its reputation as one of the go-to record stores for jazz, hip hop, electronic, and audiophile vinyl records in Singapore and the region. In 2019, TAV marked its foray as an independent record label with the establishment of TAV Records, which has, to date, released four records by Singaporean artists.

HOW DID YOU GET INTO YOUR INDUSTRY AND WHAT MOTIVATED YOU TO GET INTO IT?

I got into the industry by chance, actually. While I was always very passionate about music and collecting vinyl records growing up and did always harbor the goal of starting a music store at some point, my career was always focused, and still is, in the finance industry. It was by chance that my co-founder of TAV, Eugene Ow-Young, approached me to co-start a record store with him, with a view that I could help curate the jazz, hip hop, and electronic collections – and that was how TAV was born. I definitely saw that there was an opportunity to start a store focused on bringing jazz, hip hop, and electronic records to the audience in Singapore, as most of the other stores tended to focus on rock and pop music – I was motivated to bring in more interesting sounds to the audience in Singapore.

WHAT IS YOUR FAVORITE THING ABOUT YOUR JOB AND WHAT MADE YOU FALL IN LOVE WITH WHAT YOU DO?

My favorite thing about running TAV is the opportunity to keep discovering new and interesting sounds – there is just so much good music out there. I also enjoy working and meeting with like-minded music aficionados and introducing customers to good music and analog culture. It is also amazing that with TAV Records, we can now help promote Singapore artists via the vinyl record medium.

WHAT ADVICE DO YOU HAVE FOR SOMEONE WANTING TO PURSUE THIS CAREER?

I think it's important to know what you like and are passionate about and develop something you want to be good at.

> **" Running a business (or growing a career) is a marathon, not a sprint – and finding something you are passionate about will keep you going, especially when times get tough.**

At the same time, it is also important to understand the economics and addressable market of the industry one is in. For example, it would not make sense to open more than one physical store in Singapore as Singapore is an extremely small market with a finite number of customers willing to pay for pricey jazz records! I learned a lot about the trade, speaking to store owners when I was a customer and collector. One of the owners I learned a lot from is Mr. Alagiri Alagirisamy, who runs a store called *For The Record*. He specializes in rock and jazz original presses. His long-standing passion for great music and well-made presses has truly been inspiring for me.

WHAT IS SOMETHING YOU WISH MORE PEOPLE KNEW ABOUT THE WORK YOU DO?

Having a good team of people to work with really helps – you can't run a business well without good people. I have two long-standing employees with whom I have worked over the years. Leon and Nick, who are part of the sales and marketing team, help run the store's day-to-day operations and marketing and are responsible for curating the releases for TAV Records. TAV has also been very fortunate throughout the years to have had an amazing trove of young and energetic part-timers who help push the music we sell and promote analog culture in this part of the world.

Claudia Wilson | Pure Vinyl Records
Owner
Brixton, UK

Born, raised, and still living in Brixton, Claudia owns and operates Pure Vinyl Record Shop. She is a mother and grandmother and runs a busy home as well. She hears that she is the first black woman of Caribbean origin to own a record shop in London in recent times. Claudia sells records of all genres but mostly Soul, Funk, Jazz, and Reggae. The store stocks mostly used vinyl, but they also stock a small selection of new material, particularly from local musicians.

HOW DID YOU GET INTO YOUR INDUSTRY AND WHAT MOTIVATED YOU TO GET INTO IT?

I played records as a young child, as did most people of my generation, but I never stopped. I was a DJ for many years; before and during those years, I brought up my children and worked other jobs. I played music at many venues and had a residency at the legendary Mango Landin' Bar. With a family to keep, it was always difficult, but I wanted to buy records, so I began to sell some of the ones I already had and set up my first record stall. It has been a long road since, but I now have my own shop.

WHAT IS YOUR FAVORITE THING ABOUT YOUR JOB AND WHAT MADE YOU FALL IN LOVE WITH WHAT YOU DO?

My love of music came at a very young age. I was born in the middle of Brixton, where music was all around when record shops were part of all our lives. Record shops then allowed us to share music; for me, this is the best thing about having a shop of my own now. Owning and running a record shop is something I have wanted to do for a very long time, and though the work is immense, I love discovering new music. Using music to soothe, entertain, inform, change my mood, dance, sing, and everything else that music does really helps me to love my job.

WHAT ADVICE DO YOU HAVE FOR SOMEONE WANTING TO PURSUE THIS CAREER?

I hope that in the coming years, there will be many more women in the record industry and recognise a lack of black women like myself within the industry.

> " As women, our voices have been heard in songs forever, but now more women have shown they can be the best DJs, producers, musicians, and engineers and do much of the running of the show.

It would be great if we could further represent our voices with more record labels, shops, and radio stations owned and run by women.

WHAT IS SOMETHING YOU WISH MORE PEOPLE KNEW ABOUT THE WORK YOU DO?

Loads of hard graft! Not just the lifting of heavy boxes of records. Our shop is a community hub; we support and represent the local community in many ways. For the last six years, we have offered work experience to children from local schools. We run music groups for writing and learning, and we hosted the amazing Straight Pocket Jam nights. We offer space for local meetings and have even had film nights.

Our regular "Open Deck Nights" have been going for over ten years, since the days of the Mango Landin, and are now busier than they have ever been. We invite our customers to come down to the shop and bring their records. It is a great way to give back to our customers. We say, "Bring your records, bring a friend, bring a drink, and bring your dancing shoes!"

Shirani Rea | Peaches Records

Owner

New Orleans, LA USA

Shirani is the owner of Peaches Records, a family owned and locally operated record store since 1975. It is a cultural center caring for local musicians, painters, candle makers, jewelers, authors, and chefs. Shirani and her team are working to cultivate anyone with creative gifts and talents and help preserve the culture of New Orleans.

HOW DID YOU GET INTO YOUR INDUSTRY AND WHAT MOTIVATED YOU TO GET INTO IT?

I love people and have always had a deep passion for music.

WHAT IS YOUR FAVORITE THING ABOUT YOUR JOB AND WHAT MADE YOU FALL IN LOVE WITH WHAT YOU DO?

Music is all about love, and we are here to share that love.

> " Music is scientifically proven to heal the mind, body, and soul; we are healers.

WHAT ADVICE DO YOU HAVE FOR SOMEONE WANTING TO PURSUE THIS CAREER?

Whatever you do, it must be from your heart. It is not a business about making millions; you must truly love it. Today, I would not be able to do this without my beautiful children, Lillie and Lee, who run the business together with me.

WHAT IS SOMETHING YOU WISH MORE PEOPLE KNEW ABOUT THE WORK YOU DO?

A portion of what we take in goes to St. Jude's Mission to help the homeless and the needy, to give them a hand-up instead of a handout.

We built this store as a Wonderland for music lovers; please come share the love with us.

CHAPTER SEVEN

LATHE CUTTING

When you hear the word lathe, you may think of machinery that rotates materials like wood or metal for shaping. In this context, it's much the same idea, a machine that spins plastic discs so we can cut the sound into them. A finished **lathe-cut disc** is, in fact, a playable plastic record. If you've seen an artist selling a bespoke short run, say 20 limited copies of a release, the chances of it being a lathe cut are high. Lathe cuts use a similar process to what we discussed in Chapter 1. However, the material and steps to a completed product are very different. Due to their bespoke nature, they are typically more limited and expensive. But they also yield a larger potential market due to their collectible quality.

A lathe-cut record starts as a polycarbonate plastic, PETG (Polyethylene Terephthalate Glycol), or acrylic disc, similar to plexiglass. A CNC (computer numerical control) machine cuts the blank material to a 7", 10", or 12" size reflecting your finished record. Since the plastic is cut to size, the blank can also come in squares or odd shapes other than a circle, as well as additional sizes like 5" or 8" adding to collectability. That said, there are fewer options for colors and effects. However, as you'll see, many of the lathe cutters we've highlighted are pushing the boundaries in shapes and materials of what we commonly see and know to be possible with the medium.

Like the lacquer cut, a lathe cut is created in real-time, allowing the engineer to adjust for the best playback as they go. The engineer places a blank plastic disc onto the record lathe. The cutting head works similarly to a speaker, but rather than using a paper cone to move the vibrations of the sound wave through the air, it focuses the vibrations through an armature and down to the tip of the cutting stylus (needle) carving into the plastic, as the disc moves beneath it, creating a groove. Whether cutting into a lacquer or one of these polycarbonate blank discs, the same audio principles apply, including the type of file format and input from audio programs like ProTools. The difference is that a lacquer has a coating too soft to stand up to repeated plays, so it goes through the electroplating process (as discussed in Chapter 2). By contrast, the lathe cut is ready for labeling and packaging after cutting. Since each disc is a playable copy, the engineer must repeat the cutting process for each record over and over in real-time. The real-time nature of this process is why the format lends itself to short runs. Distinguishing itself significantly from the requirements of a pressing plant, orders for lathe-cut records can be as small as one copy, up to about 100 pieces. In addition to smaller order sizes, this process commonly has a faster turnaround time because the additional electroplating steps required to manufacture on a press are unnecessary.

The cutter creates each piece one at a time; this means all other manipulation of the end product occurs the same way. The application of center labels, package assembly, blank side etching, and screen printing are all completed by hand, one by one. While labels in record pressing are not stickers, here they can be, and there are more opportunities to create a custom, one-of-a-kind piece. An option like this makes it a perfect way for bands, artists, and labels to do specialty items for a release or a short run to gauge interest in a new single.

With all the variations discussed and common knowledge around how music is released today, this may seem like a newer technology. However, lathe-cut records have been around for quite some time, and many of the lathes in use today are restored vintage mono record lathes. Lathe-cut records go back to 1917, but their story got even more interesting during the cold war era of the late 1940s in Eastern Europe. During this time of extreme soviet oppression, all non-state-sanctioned material was outlawed, including western music and local heritage arts like literature, poetry, film, and folk music. There are fascinating stories of contraband Russian bootleg lathe-cut recordings flourishing on the black market during the years after World War II. These highly collectible pieces of history, often referred to as "**Bone Music**," were cut into recycled x-ray film salvaged from the trash behind hospitals.

You might ask yourself, "But how do they sound?!" because that is what matters. Where x-ray bootleg sound quality was at times almost inaudible, today's modern lathe-cut records can shine. However, they are not necessarily equal to a pressed record. A record lathe is a tool, so naturally, there will be variation based on the operator's knowledge vs. the quality and tools they use. Therefore, lathe-cut records are not monolithic and will sound different based on the cutter's equipment, method, and skill.

Additionally, there are two different methods of making a modern lathe-cut record. **Mono embossing** scratches the grooves into the plastic and tends to be lower fidelity, lower volume, with shallower grooves. This method is usually less expensive and tends to be used more often as band merch because of the price point. Records that are embossed by an experienced cutter sound better, traditionally, and can play on a low-end turntable but cannot be manipulated in the same way as a regular record. Then, there is **stereo diamond** cutting which removes material from the plastic, resulting in a higher fidelity cut with a deeper groove. However, this method comes at a higher price point and is often used for one-offs and short-run specialty releases. In each case, a tonearm adjustment on your turntable may be necessary for the best sound due to these variations in mono vs. stereo cuts or groove depth and shallow vs. deeper grooves.

To some artists, the lo-fi or mid-fidelity sound is part of the reason for choosing a lathe cut, unlike the vinyl audiophile market. The best way to get what you want from this format is to talk with the engineer or search out their work since it is a more intimate manufacturing process than a pressing plant.

Robyn Raymond | Red Spade Records
Owner
Calgary, Canada

Robyn Raymond at Red Spade Records creates bespoke, handmade, lathe-cut records and is currently the only operating female record cutter in Canada. Her lifelong passion for music and vinyl records inspired her to create the company, and her keen eye for detail makes her work impeccable and unlike any other. Bringing artists' music to life in the physical form is Raymond's true calling. She is adamant about making vinyl accessible and co-hosts the Women in Vinyl podcast, as well as taking any opportunity to speak about her non-traditional path into the world of mastering and helping to navigate the various ways of making physical music.

HOW DID YOU GET INTO YOUR INDUSTRY AND WHAT MOTIVATED YOU TO GET INTO IT?

I got to where I am now by way of a lifelong infatuation with records. My family moved around a lot, and the one thing that was always constant was music, particularly records, which remained a major part of my life. I worked my way through the ranks in live shows from 2005 to 2012, doing marketing and other related duties for everything from club shows to arenas and festivals. I got well-acquainted with everything that goes into making a big show or tour happen, but eventually went back to school for more studies in human performance – my original field of study. I then found my way to working for the Canadian Olympic bobsled, skeleton, and hockey teams as a manual therapist from 2012 to 2017. After the conclusion of the 2016–2017 season and a month in Korea doing the Olympic homologation, my contract wasn't renewed, and I was still trying to figure out what I would do next.

During a trip to Edmonton to see Metallica with an old bandmate, we lamented over our lives and dissatisfaction with our less-than-rockstar existence. He suggested I get in touch with the people who run Canada Boy Vinyl, a now-defunct Calgary-based record-pressing plant, to see if there was any work I could do with my knowledge of vinyl. Through that conversation, I discovered there are many ways to make a record, which I knew, but not to the degree I do now!

I connected with Calgary's Inner Ocean Records, whose owner (Cory Giordano) turned me onto lathe cutting. Even though I felt out of my depth, having no formal education in audio, but confident in my ability to learn by doing, I felt I had the necessary skills and connections to make a career in vinyl cutting.

WHAT IS YOUR FAVORITE THING ABOUT YOUR JOB AND WHAT MADE YOU FALL IN LOVE WITH WHAT YOU DO?

I *love* doing the "*feels*" records. The one-offs include the "song I wrote for my wife for our anniversary" or the last voicemail that your loved one left you. Nothing beats putting a physical copy of an album into the musician's hands, though. There's something about transforming digital music into a physical thing they can hold in their hands.

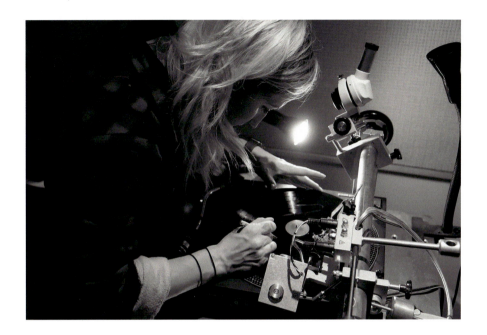

WHAT ADVICE DO YOU HAVE FOR SOMEONE WANTING TO PURSUE THIS CAREER?

Cutting records is awesome, but it's not for everyone! The most beneficial thing for cutting records is learning to be a great mastering engineer. Then, go to a lathe-cut camp in sunny Tuscon with the incredible Mike Dixon, and see if this might be for you! This is the one area where you need to have the gear, so being able to try it out first is the best course of action.

> **"** Luckily, cutting records and mastering is a very welcoming club. So making friends and allies in this niche is easy, fun, and will also be super helpful.

WHAT IS SOMETHING YOU WISH MORE PEOPLE KNEW ABOUT THE WORK YOU DO?

That I care about your records as much OR MORE than you do. I spend probably just as much time working on them. I'm a one-woman show; I do everything from the beginning to the end. Lathe-cut records sometimes play a bit differently than pressed records, so your playback system may require some adjustments. But that's totally normal. I guarantee that I've done all the things to ensure they are the best that I can possibly make them.

Bailey Moses | Hocus Bogus Records
Founder
Los Angeles, CA USA

Bailey's partner likes to call them a peddler of novelty vinyl and wares, which they think is a funny but apt description. They run a boutique record label and lathe-cutting service called Hocus Bogus Records, which primarily focuses on creating wacky and unique lathe-cut records. Bailey collaborates with different artists and record labels to make their weird record dreams come to life. You want a record with real birthday candles coming out of it that plays Happy Birthday? Sure! You want them to turn your favorite childhood DVD into your band's new single? Let's do it!

HOW DID YOU GET INTO YOUR INDUSTRY AND WHAT MOTIVATED YOU TO GET INTO IT?

I stumbled upon record-making somewhat accidentally. I met the lathe cut wizard Mike Dixon while I was in college and asked if he needed any help sweeping or cleaning up around his studio. I showed up for my first day and, to my surprise, got a crash course on how to operate a Presto 6N. I immediately fell in love with it and haven't stopped since.

WHAT IS YOUR FAVORITE THING ABOUT YOUR JOB AND WHAT MADE YOU FALL IN LOVE WITH WHAT YOU DO?

> " I love pushing the boundaries of what a record can be and seeing people's reactions when they realize this weird, monstrous thing I've pulled out of my garage can be played on their fancy turntable.

Maybe it's a reach, but I like to think of my lathes as sort of a punk gesture–a f**k you to the mainstream cis male-dominated record world. I'm still a vinyl collector at heart, but I also love creating records that go against audiophile culture and push up against the idea of being pristine.

WHAT ADVICE DO YOU HAVE FOR SOMEONE WANTING TO PURSUE THIS CAREER?

If you're interested in getting into lathe cutting, I'd definitely recommend checking out Mike Dixon's lathe cut camp. I've helped him do a few camps before, and it's a really great, holistic view of what you're really getting into. Most people I meet tend to think of making records as a bit more glamorous than it really is. The truth is, it's a lot of work, and using a machine from the 1920s can make you want to rip your hair out. It's not for everyone, but if it IS for you, you'll have the time of your life!

WHAT IS SOMETHING YOU WISH MORE PEOPLE KNEW ABOUT THE WORK YOU DO?

It has become less of an issue now that lathe cuts are growing in popularity, but the big thing I wish everyone knew is that lathe cuts and pressed records are not the same thing. Lathe cuts are made on plastic, in real-time, on a machine from the 1920s, by hand. They are never going to sound the same as a pressed record. They are not vinyl. Once people are aware of all of this, it opens them up to the wonderful world of possibilities that lathe cuts can offer.

Tasha Trigger | Lathe to the Grave
Owner
Cardiff, UK

Tasha is the owner of Lathe to the Grave, a company she started in December 2017 primarily as a way to help her friends release lathe-cut vinyl records, as there weren't a lot of options out there at the time offering short-run, quick turnaround services.

HOW DID YOU GET INTO YOUR INDUSTRY AND WHAT MOTIVATED YOU TO GET INTO IT?

My partner has been making music all his life, attending university as a mature student to make music his career rather than just a hobby. I was inspired by his commitment to completely changing his life. I had worked in the public sector for the best part of a decade with no opportunity for career progression.

As he was approaching the end of his degree, we were discussing how difficult it is to release vinyl as an independent artist over a few pints in the Dancing Man pub in Southampton, UK. I started researching how records were made, from lathes to pressing plants. I combed the internet for information on how the process worked, learning how to master music specifically for vinyl, and, most importantly, hunting down a lathe to purchase.

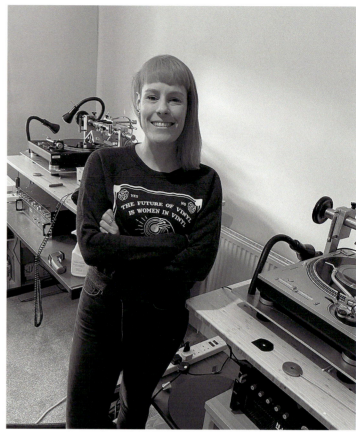

> **❝ It took almost a year from this decision to obtain a lathe but finally picking it up was the best feeling and a little adventure!**

It was another six months before I felt comfortable that the records I was making were good enough quality for me to start offering them as a product.

WHAT IS YOUR FAVORITE THING ABOUT YOUR JOB AND WHAT MADE YOU FALL IN LOVE WITH WHAT YOU DO?

I love that every week is different. We cut all sorts of music from all kinds of people. Some weeks it might be a big run of albums for a folk artist via an agency; the next week, it is a run of 6 copies for a super independent artist in a very niche genre. I get to chat with new people every week about making new and unusual vinyl.

It also helps that I am now self-employed and not having to work for someone else; I have a better work-life balance. I can exercise in the morning, sit and have breakfast with my partner before we go in different directions at the top of the stairs, him to his studio, and me to my workshop.

WHAT ADVICE DO YOU HAVE FOR SOMEONE WANTING TO PURSUE THIS CAREER?

Do your research extensively before taking the leap. The forum Lathe Trolls is my go-to whenever I have any issues. Also, pick up textbooks, Larry Boden's *Basic Disc Mastering* is always on hand in my workshop. Lathe cutting and vinyl have been around since long before the internet, and some of the best resources are good old-fashioned books.

It is an extremely labor-intensive and expensive hobby to pursue. Even once you feel comfortable about the process and using your equipment, there are lots of expensive, easily breakable parts. It is definitely a labor of love, and you really have to be all in for it to work.

WHAT IS SOMETHING YOU WISH MORE PEOPLE KNEW ABOUT THE WORK YOU DO?

How long it takes! I think everyone has seen videos of pressing plants where the puck gets squashed between the stampers for a few seconds, and voila, a record is made. Lathe cuts happen in real-time, sometimes even at half speed, so when someone asks for a run of 50 × 12" records, I don't think they understand that it will be 50 hours of continuous work, plus preparation time, quality checks, labeling, sleeving, packing, etc. It really is a labor of love.

Emily Nobumoto | Leesta Vall Sound Recordings

Co-owner & Director

Brooklyn, NY USA

Emily Nobumoto is the co-owner of Leesta Vall Sound Recordings, a niche record label based in Brooklyn, NY. Since graduating from Purchase College in 2019, she has worked with over 2,000 artists via the label's main initiative, Direct-to-Vinyl Live Sessions. Hosted at their Brooklyn studio, these sessions are live performances lathe-cut directly to 7" discs. No tracking or overdubs of any kind, just a song played live, mixed, and mastered on the fly, and cut in real-time to a limited edition vinyl record. Each record produced is a completely unique, one-of-a-kind musical artifact, never to be duplicated or shared digitally among the masses.

HOW DID YOU GET INTO YOUR INDUSTRY AND WHAT MOTIVATED YOU TO GET INTO IT?

I grew up heavily invested in music and have always been driven to work in the music industry. I've explored everything from performance to artist management to art & entertainment law. Ultimately, I got my degree in Arts Management, which led me to an internship opportunity at Leesta Vall. I dug my heels in at this company and have been working to grow it ever since!

Seeing vinyl making a comeback since the late 2000s has been so inspiring.

> **The ritual of making and listening to vinyl is a positive, community-building experience that I've wanted to be a part of for as long as I can remember.**

Vinyl has quite literally become the anti-MP3 – a sort of symbol of resistance against the digitalization of music. We live in a fast-paced, digital world that thrives on instant gratification, and vinyl records are the complete opposite. They offer a tangible, multi-sensory experience for the listener that goes way beyond pressing play on an MP3. Spinning vinyl forces you to slow down and truly experience the music in the way it was intended.

WHAT IS YOUR FAVORITE THING ABOUT YOUR JOB AND WHAT MADE YOU FALL IN LOVE WITH WHAT YOU DO?

Leesta Vall has the ability to work with almost any artist at any stage in their career, no matter their level of exposure. We offer the opportunity for independent artists to make their own vinyl records without the high cost and long wait time of a typical pressed release. Our mission also allows a direct, one-to-one connection between the artist and the listener. At the end of the day, seeing the magic happen in the studio when an artist records a song for their friend, their mom, or their biggest fan, makes it all worth it – knowing that person will cherish their record and be transported to a specific moment in time when it was recorded by someone they love.

WHAT ADVICE DO YOU HAVE FOR SOMEONE WANTING TO PURSUE THIS CAREER?

Don't be afraid to start something you have no experience in. The lathe-cutting community is extremely accepting and understanding. We're not working with an exact science here; almost everything is trial and error. The only way you learn is by making mistakes! Working with vintage equipment takes a lot of practice and patience, but the experts and mentors in this industry would love nothing more than to pass on their craft to younger generations. And take the necessary time to do your best work. Anything less isn't worth it.

WHAT IS SOMETHING YOU WISH MORE PEOPLE KNEW ABOUT THE WORK YOU DO?

Lathe cuts are sometimes misunderstood and expected to have the same audiophile quality as a pressed record. But since they're made differently, they sound different. Contemporary vinyl consumers are used to hearing the hi-fi quality of pressed releases, so it's important to prioritize minimizing surface noise on lathe cuts while still preserving the warm, lo-fi heart of the record that we know and love.

Oihane Follones | Lathesville

Co-owner

Berlin, Germany

Oihane is the co-owner of Lathesville, a mastering and lathe-cutting studio in Berlin. Born in Basque Country, Spain, she moved to London when she was 18 to work at a record store, and is now located in Berlin where she runs the record store Wowsville in conjunction with Lathesville. She also handles European distribution and mail order for Slovenly Recordings. From running a lathe cut business, to working in record stores, managing and working with indie labels, DJing and playing drums in a band, she does it all.

HOW DID YOU GET INTO YOUR INDUSTRY AND WHAT MOTIVATED YOU TO GET INTO IT?

I've been working with vinyl for the last sixteen years. First in London at a 50s specialized record store, then ten years ago, I moved to Berlin to run a record store called Wowsville. I also do the European distribution and mail order of Slovenly Recordings. Since January 2022, I have been the co-owner of Lathesville, a mastering and lacquer-cutting studio in Berlin.

During the pandemic, while the record store was closed, the idea of setting up a pressing plant started brewing. We bought a lathe machine, and in 3 months, we got a great team together with the help of some fine folks. We started cutting at the end of March, upgraded our lathe in May, and have been rolling since then.

The plan is little by little to get a pressing plant so we can offer all the steps of the vinyl manufacturing process. So far, we have the cutting studio in Berlin, and we are starting to set up the rest of the production in Valencia, Spain.

VINYL RECORD CUTTING LATHE MACHINE
NEUMANN VMS

WHAT IS YOUR FAVORITE THING ABOUT YOUR JOB AND WHAT MADE YOU FALL IN LOVE WITH WHAT YOU DO?

Overall, music is the gift that keeps on giving, not only on the emotional and therapeutic level but also because of the amazing people I have met, thanks to it.

That said, there are a few aspects of the job we do that I love. First and foremost is to be able to understand sound with more than one sense. Not only by hearing it but also by seeing it. How an audio recording is transformed into grooves is fascinating and to be able to look at those grooves through a microscope and understand how that translates into music is mind-blowing.

I also like the combination of analog and digital technology. We use old techniques executed with a combination of old machines and a set of state-of-the-art upgrades and software. Our lathe machine is from 1966; she is an old lady, but thanks to a new pitch and software developed by one of our engineers, we are able to push what the lathe can do to the next level.

WHAT ADVICE DO YOU HAVE FOR SOMEONE WANTING TO PURSUE THIS CAREER?

If you are passionate about it, just go for it, don't let anyone tell you you can't do it.

> **" Dedication and perseverance will take you very far; you can learn the rest along the way.**

If you are a woman, don't be afraid to bulldoze your way through this male-dominated industry.

These are skills that can't be learned in a school, and there is a lot of gatekeeping in the field; therefore, in my opinion, those skills need to be passed on to the next generation to ensure that vinyl survives. I want to make a special shoutout to all the people back in the 90s when vinyl sales plummeted, and all the majors switched to CDs; they kept vinyl alive. All the independent labels, underground music freaks, and DIY nerds that invented ways to improve old machines made vinyl's survival possible. I will be forever grateful to people like Flo Kaufmann, Martin Sukate, and Helmut Erler for all the hard work they have done.

I also get inspired by other women; I love what you are doing with "Women in Vinyl." Mary Dee Dudley, Gladys Hopkowitz, Sylvia Robinson, Linda Stein, Miriam Linna, and many more opened the way for us.

WHAT IS SOMETHING YOU WISH MORE PEOPLE KNEW ABOUT THE WORK YOU DO?

The cut is very important to the overall sound of the final product. Bands and record labels, please make sure you have a dedicated cutting engineer doing this task. We will always make sure your record sounds great and endures for a very long time.

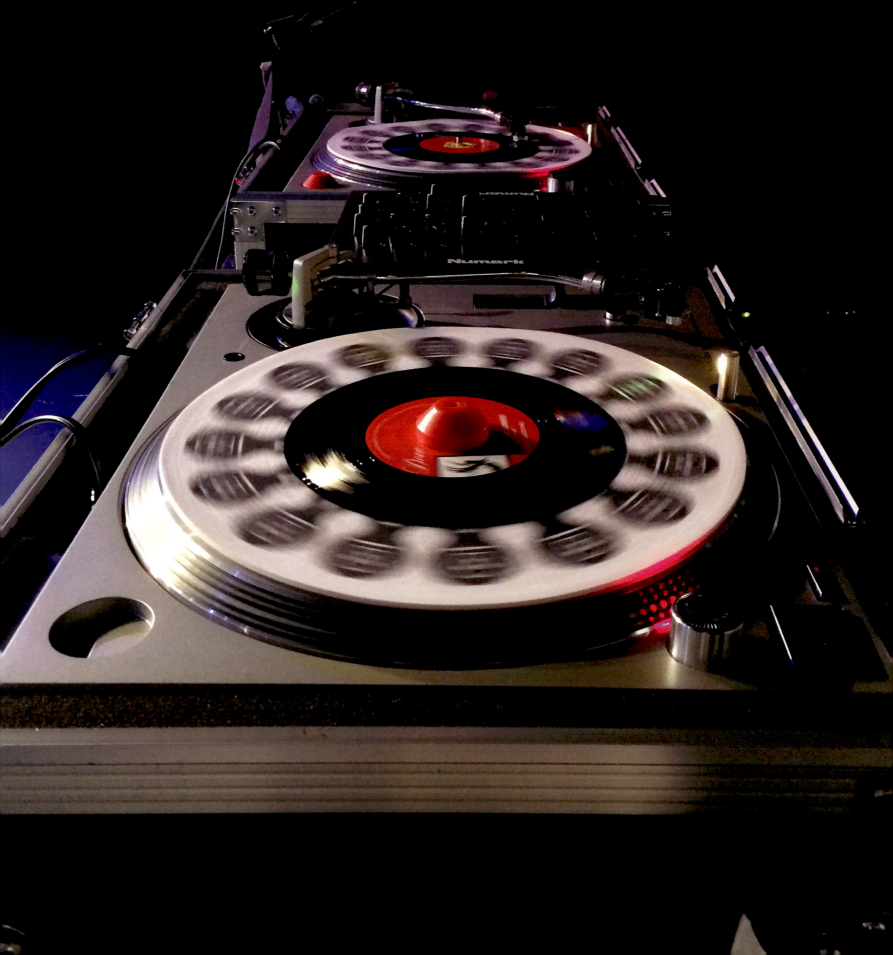

CHAPTER EIGHT

DJING

When you picture a disc jockey or DJ today, you may think of a festival stage with a digital setup controlled by a performer whose goal is to keep the crowd in a particular mood or vibe. However, before streaming and television, discovering new music was much harder than it is today. DJs playing vinyl on the radio were integral to entertainment and music discovery. The term might have been coined by radio commentator Walter Winchell in 1935, referring to Martin Block, who created memorable experiences for listeners during his program. There is also speculation, however, that Jack Kapp, a record executive with Brunswick Records and founder of American Decca Records, coined the term in 1940, saying presenters were "record jockeys"; either way, the term stuck.

Several types of DJs have forged unique niches since radio's heyday, such as club DJs, mobile DJs, and turntablists. But generally, all DJs share the basic definition of someone that mixes recorded music, typically specializing in a particular genre(s). While there is no doubt that radio DJs were creating a mood and curated playlist, there is also the social and dance aspect to this part of the industry (mobile and club DJs) that is integral to the growth of DJing. Playing vinyl for dancing and parties was popularized with the marketing of home turntables in the late 19th Century. But the first dance party to have a DJ occurred in 1943 in Otley, England. Playing jazz records, radio DJ Jimmy Savile launched the world's first DJ dance gig. He also gets credit as the first to use two turntables for continuous play in 1947, which has become a staple of vinyl DJ culture.

Throughout the 1950s, event promoters began to call themselves "deejays" in parts of Jamaica, hosting dancehall and reggae parties in the streets. The DJs were referred to as the "selector" and played music through a PA system for all to enjoy. The success of these parties caught on, and the idea expanded elsewhere in the world and eventually evolved into paid events. In the late 1960s to early 1970s, this became known as the Jamaican sound system culture; you can think of it as the advent of mobile DJs.

Meanwhile, nightclub and discotheque culture grew in Europe during the 1960s, with the United States not far behind. Long music-filled nights led by DJs began popping up everywhere. During this period, we also see performers adding "DJ" in front of a creative stage name or pseudonym to represent their set and make themselves more distinguishable and marketable.

The 1970s is when we see DJs emerge as we recognize them in pop culture today. We start to see the merging of hip-hop music, DJing, and urban culture with promoters and MCing (or rapping) to combine mixing techniques, song samples, percussion breaks, and basslines from previously recorded music. This scene took the DJ career from curating a playlist to a true vinyl art form. During this time, DJ Francis Grasso popularized beat-matching, the mixing or blending of songs with a seamless transition between tracks with matching beats or tempos, becoming the foundation of modern club DJs.

Around the same time, DJ Kool Herc began to experiment with techniques like mixing between identical albums to extend breaks. Grandmaster Flash pioneered hip-hop DJing, scratching, and mixing. Scratching involves moving the record back and forth under the needle to produce a rhythmic sound. The dual turntable setup they used with headphones, a mixer, and a crossfader allows the DJ to cue and play two records simultaneously, manipulating them using these techniques to line up, blend, and overlay tracks creating unique and unforgettable mixes. Turntablists, or battle DJs, take all these techniques and equipment even further to manipulate recorded sound into completely new music using their turntables as an instrument, participating in contests and competitions against each other.

It is essential to take a moment and focus on the technology at the very center of DJing culture, the turntable. DJing has become what it is today due to the creation of the direct-drive turntable in 1969 by engineer Shuichi Obata at Matsushita (now Panasonic) in Osaka, Japan. Belt-drive turntables to this day don't provide the torque to allow manipulation of the record. Their slow start-up to 33 1/3 rpm isn't conducive to the pace DJs require, and they were prone to breakage under this type of use. Matsushita's direct drive turntable, the SP-10, was the first in the market's most highly regarded turntable series Technics. In 1972, when Technics started making the SL-1200, it was a match made in DJ heaven and what has become the quintessential DJ turntable.

Although much beloved, records are heavy and can be cumbersome to lug around town to gigs. Always capitalizing on the newest technology, the "traditional DJ setup" evolved in the 1980s, transitioning to include cassettes and, later in the 1990s, onward to digital. Some technological advancements have allowed for a consistent experience for the audience and performer with software programs like Serato or turntable emulating hardware like CDJs. These digital media players enable DJs to use control wheels for pitch control of digital files. Providing more music at the DJ's fingertips with the press of a button allows for more flexible setlists eliminating the challenge of being locked into the records brought with them. However, some still agree that vinyl is the proper way to curate, share, and move a room to music.

Colleen "Cosmo" Murphy | Classic Album Sundays

Founder & DJ

The World

Colleen is a musical curator, DJ, radio host, producer, and educator. She shares music through her radio shows and DJ sets at clubs, festivals, and parties around the globe. Colleen also makes music as both a producer and remixer. In 2010 she founded Classic Album Sundays, an event and content platform that tells the stories behind our favorite albums and hosts album-listening events worldwide. Music is her life.

HOW DID YOU GET INTO YOUR INDUSTRY AND WHAT MOTIVATED YOU TO GET INTO IT?

Music has been an obsession since I got my first transistor radio at the age of 7, and I spent hours scrolling up and down the bandwidth discovering new songs and sounds. A few years later, I began broadcasting on our 10-watt high school radio station and then started working at a record shop after school and began record collecting. I had an open mind and an open ear and bought records from all ends of the spectrum. Then I moved on to college radio, a powerful force in the mid-80s as we broke bands and played records not supported on commercial radio. I found that putting my head down and getting on with the work opened up other doors and opportunities like Program Director. That role led me to a stint as a radio host in Japan and then a full-time job hosting, writing, and producing syndicated radio shows, where I had the chance to interview some of my favorite artists and bands.

Around this time, I started going to parties and clubs in New York City and landed at my friend David Mancuso's party, The Loft, another turning point that brought me into a new musical world. I began DJing first locally, then throughout the country, and finally internationally whilst holding down other jobs in yet another record shop, at a record label, and with a DJ record pool. A big part of DJ culture is also making music. I began writing and producing my own songs and remixing other artists and had an especially productive run in the studio after having my daughter.

In 2010 I founded Classic Album Sundays, which started as a monthly album-listening event in my friend's pub. It has since grown to host events around the world, sometimes featuring the artists themselves, and a website that features artist interview videos, podcasts, playlists, and blogs all about the elixir of life: music.

WHAT IS YOUR FAVORITE THING ABOUT YOUR JOB AND WHAT MADE YOU FALL IN LOVE WITH WHAT YOU DO?

The joy of musical storytelling and sharing the healing force of music.

WHAT ADVICE DO YOU HAVE FOR SOMEONE WANTING TO PURSUE THIS CAREER?

Volunteer, apprentice, or work for smaller businesses to gain experience in all aspects of your vocation.

> **“ Do your best and put your heart into it; it will lead to other opportunities.**

Eventually, create your own thing. It's more fun that way.

WHAT IS SOMETHING YOU WISH MORE PEOPLE KNEW ABOUT THE WORK YOU DO?

Because I have worn many different musical hats, I often found it difficult to bring it all together and often divorced various aspects of my career from one another. I was "Cosmo" as a DJ and producer, and "Colleen Murphy" as Classic Album Sundays' founder and host. It took me a while to realize there were always two common elements: music and myself. I have since brought it all together as a holistic vocation. I am now allowing myself to enjoy being a multi-faceted music person who has created her own musical world.

Misty Fujii | DJ Misty

Vinyl DJ

Osaka, Japan

Originally from Toronto, Canada, Misty moved to Osaka, Japan, where she and her husband are raising their son. Misty spent the last two decades selecting the best in Rock n' Roll, Soul, Funk, R&B, Punk, and everything in between at bars, clubs, corporate/private events, and across the airwaves in Toronto, NYC, LA, Mexico City, Tokyo, Kyoto, and Osaka. Today she predominantly plays vintage R&B and Soul music in all its forms on vinyl, online on Night Beat Radio, and at venues in Japan. She's a proud founding member of the Toronto Soul Club. Misty's husband runs a record store called Night Beat Records, which she loves being part of. In her free time, she is also blogging about her experience becoming a parent in a foreign country at: www.fujiiyamamama.com

HOW DID YOU GET INTO YOUR INDUSTRY AND WHAT MOTIVATED YOU TO GET INTO IT?

I've been around music my whole life with a musician father and a house full of records growing up. I started DJing for fun in college so my friends and I could hear the music we liked when we went out, and it just exploded from there.

Another reason I started DJ'ing was to help with my social anxiety. Being the DJ gave me a reason and a purpose to go out, and whenever I felt overwhelmed, I could "hide" behind the music in a DJ booth. I'm grateful to have worked through my anxiety (though it still crops up from time to time!), but DJ'ing was a way to support me through that.

WHAT IS YOUR FAVORITE THING ABOUT YOUR JOB AND WHAT MADE YOU FALL IN LOVE WITH WHAT YOU DO?

One of my favorite things about DJ'ing is traveling to DJ and meeting other music fans. I fell in love with DJ'ing when I could make people dance and see how happy they were. Watching a dancefloor go wild is an addictive rush I've never stopped chasing! Since I started, I've been lucky enough to provide the soundtrack for people's most memorable nights and events, and the love of doing that will never be lost on me.

WHAT ADVICE DO YOU HAVE FOR SOMEONE WANTING TO PURSUE THIS CAREER?

The best first step for anyone who wants to get into vinyl collecting and DJ'ing is getting to know other collectors and sellers and immersing themselves in their city's music scene. Work with other local DJs and promoters, and go for it!

Another suggestion is to immerse yourself in the music you're playing. Learn about the artists you like and why they made the music they did. Nothing irks me more than DJs or collectors of 60s soul music saying, "don't bring politics into music," when that time was incredibly political for Black artists. If you love the music but can't appreciate the social climate that went into making the music, then you aren't genuinely listening.

> **❝** Something I had to learn was how important it is to leave ego at the door. As much as a DJ's job is to highlight music that people may not have heard, we're also there to connect with people. That means playing to your

crowd and ensuring everyone is having a good time – not just being educated. The best DJs are the ones who can seamlessly do both. Finally, have fun! DJing is one of the best jobs in the whole world.

I've been inspired and have learned so much from the other members of Toronto Soul Club, who are all incredible DJs and entertainers. I'm also continuously inspired by women who DJ and work to build up other women and non-binary DJs.

WHAT IS SOMETHING YOU WISH MORE PEOPLE KNEW ABOUT THE WORK YOU DO?

I wish people knew how heavy it is to carry a crate of records! Just kidding, I wish more people knew that a lot of genuine love and passion goes into what we do. It takes years to collect enough records to build DJ sets, which requires serious dedication. I'm so passionate about what I do that I DJ'ed up until a few days before my baby was born! And now I'm back at it because I can't stay away from the music for long.

I want people to know that women and non-binary DJs know our s!?t! We know music and what we're doing just as much as any guys do. It doesn't matter what we're wearing or how we look. I also want people, especially women and non-binary people, to know how much fun it is. Hopefully, more women and non-binary people will continue to carry the torch that is playing these wax platters to people for decades to come.

Monalisa Murray | DJ Monalisa
Vinyl DJ
Los Angeles, CA USA

Monalisa is a DJ based in Southern California as well as the Used Product Buyer for Amoeba Music in Hollywood. She has been associated with quality music and events in Los Angeles and across the United States for over 20 years as a music promoter and fan before taking to the turntables over 16 years ago. Her music format is "anything with a groove," ranging from soul, funk, hip hop, rock, pop, jazz-funk, world, and less familiar gems of all genres. Monalisa is a member of KPL All-Stars, Umoja Hi-Fi Soundsystem, Ladies Of Sound, and Prism DJs.

HOW DID YOU GET INTO YOUR INDUSTRY AND WHAT MOTIVATED YOU TO GET INTO IT?

I learned how to DJ by watching my friends Vicious Lee and Mark Luv at their gigs in the early 90s, and they both gave me hands-on lessons on how to count BPMs and blend records together. Back then, I worked in the music business as a record promoter, so I would give promotional records to the club and radio DJs in the city and mess around over the years whenever I came across turntables. I didn't start playing out seriously until 2006. My friend DJ Dusk passed away that year, and he would always play music off the path of the usual DJ no-brainer playlist. We'd always have great conversations about those uncommon records, and after he passed, I felt the need to play those kinds of records and continue those conversations with others.

One day I was at Spinderella's house, and Pete Rock was there playing video games. I decided to play some of her records to pass the time, and Pete was listening. He complimented me on my selections and asked, "are you a DJ?" I said, "not really," and he said, "you should be; you're killin' it. Dope selections," and went back to his game. I picked my jaw up off the floor and hit the ground running. After putting the word out that I wanted to start playing, I was gifted a pair of turntables from one of my DJ heroes, and Spin let me borrow a mixer and a pair of needles. I started getting my records in order and playing early opening bar and club sets, and things went from there.

I started working at Amoeba Music in 2009 as a cashier and floor person. I was promoted to used product buyer in 2014 after my co-worker and friend Kevin Fitzgerald resigned and recommended me to fill his position. I was full of anxiety when I started because I was given such a huge responsibility, and I was afraid that I would do something wrong, like paying too much or too little, but I gradually got the hang of it and am still at it eight years later. I know so much more about music, especially vinyl, because of that experience.

WHAT IS YOUR FAVORITE THING ABOUT YOUR JOB AND WHAT MADE YOU FALL IN LOVE WITH WHAT YOU DO?

My favorite thing about being a DJ is bringing joy to the listeners and fans through music and educating them about the music they may not be aware of. It's also my favorite thing about working at Amoeba – helping people find music they love and helping them to discover music that will add to the soundtrack of their lives.

WHAT ADVICE DO YOU HAVE FOR SOMEONE WANTING TO PURSUE THIS CAREER?

I advise anyone wanting to become a DJ to watch and study the DJs you admire and learn from them. I watched the Beat Junkies Crew over the years, and they are hands down some of the best DJs to ever touch turntables. I was hesitant when I started because I felt that if I wasn't going to be as good as them, then I shouldn't bother. I later realized that they were on a whole different level of DJ skill and that I didn't have to be as good as them… I could do my best and still have a great career. I am currently attending the Beat Junkie Institute Of Sound and taking the entire 3-level course because I want to sharpen my skill, and I highly recommend it to anyone in the Los Angeles area. http://www.beatjunkiesound.com

WHAT IS SOMETHING YOU WISH MORE PEOPLE KNEW ABOUT THE WORK YOU DO?

> **“ I wish more people truly understood that all DJs don't just push buttons and turn knobs.**

Those who take the craft seriously go through a lot to move crowds… we study, prepare, set the mood, tell a story, and bring people on a journey. It's so fulfilling when people trust us with the journey and let us drive.

Dana Brown | Queer Vinyl Collective
Founder & DJ
Austin, TX USA

Dana is a vinyl DJ and a lover of all things quirky. Her world revolves around collecting vinyl and building community in the local vinyl scene. As founder of Queer Vinyl Collective (QVC), an Austin-based community of queer-identifying folks, she's working to carve out more space for queer artists and performers while creating a more equitable space for the LGBTQ+ community.

HOW DID YOU GET INTO YOUR INDUSTRY AND WHAT MOTIVATED YOU TO GET INTO IT?

I'm a trans woman who came out in the early pandemic days. Once I began presenting as a woman, I found it hard to engage with vinyl collecting and the industry in general despite having a great fondness for and lifelong exposure to it. Walking into a record shop or a music store before my transition was a smooth process and a situation where I comfortably engaged with folks because I was a "male." After presenting as my true self, I began to feel like my interest and expertise were not taken seriously. I began to experience what so many femmes in the industry have endured forever, and it made me sad and more than a little angry. After lots of frustrated conversations with anyone that would listen and personally ideating on a better way, Queer Vinyl Collective started to take shape. QVC was created in direct response to this pervasive issue in the music industry. Girls, women, and frankly, most queer-identifying folks, in general, are not treated with the same level of respect that masc-identifying folks are. I created QVC to help combat this and make a real sense of cohesion in the Austin vinyl scene.

WHAT IS YOUR FAVORITE THING ABOUT YOUR JOB AND WHAT MADE YOU FALL IN LOVE WITH WHAT YOU DO?

Engaging with the community at our events is the most unique and enriching aspect of what I do and often is what keeps me going. DJing at various clubs, shops, and venues in town is fun, but it's nothing without the chance to engage with an audience and community. QVC celebrates those just growing their collections and passions for vinyl; our events are the best way to connect with those folks. I've had the pleasure of getting several friends behind the decks and seeing them blossom into really stellar DJs. I have a deep joy for getting people more connected with music, and this is no different.

WHAT ADVICE DO YOU HAVE FOR SOMEONE WANTING TO PURSUE THIS CAREER?

> Do it, scared. Don't talk yourself out of what is a good idea. Other people want to do that thing that you see as "silly."

Vinyl collecting, DJing, and community building through this conduit have been so restorative, although QVC started with nothing more than a desire to engage and create community. While it's been a lot of hard work to get it off the ground, the reward has been worth it. Honing my craft as a vinyl DJ has been a lovely new way to revisit songs that I love and have shaped my musical taste, but it's also been this rich way of discovering new artists or ones I missed the opportunity to enjoy the first time around. Vinyl DJing organically pushes you to find more tracks to weave into your sets, and I can't think of a more magnificent reason to keep exploring!

WHAT IS SOMETHING YOU WISH MORE PEOPLE KNEW ABOUT THE WORK YOU DO?

Often, folks assume DJing means you've been a collector for ages, have a massive catalog to work with, and you tote it to every gig so you can meet the moment. In reality, anyone can begin to collect records of their favorite music and become a DJ too! One of the most important things a friend told me when I first began DJing is that those minor issues like a limited catalog, choppy transitions/skipping tracks, and other DJ mishaps are just part of it. We all have those moments, which often make this creative expression more fun. It's imperfect, and that's okay. I am not the most proficient DJ by any stretch, but I have fun, and that's certainly all that matters when you're behind the table. If more people knew just how easy and exciting vinyl DJing can be, loads more would surely get involved.

DJ Honey
Vinyl DJ
Philippines

DJ Honey is a radio host, promoter, collector, and vinyl DJ living in the Philippines with experience DJing in Australia, the United States, United Kingdom, and Asia. She plays a sultry, raucous, and exciting mix of RnB, soul, jazz, and pop-yeh-yeh on original vinyl 45s from the 1950s–1970s. Honey likes celebrating femme fatales, vintage vixens, and "bad girl babes" of the era who consoled broken hearts, sang about love, promoted girl power, and inspired a generation of music lovers. In 2016 she began hosting Kiss! Kiss! Bang! Bang! on Brisbane's 4ZZZ FM, now on London's Soho Radio and Mixcloud.

HOW DID YOU GET INTO YOUR INDUSTRY AND WHAT MOTIVATED YOU TO GET INTO IT?

I started DJing after a particularly traumatic and life-changing experience. I decided not to allow fear or intimidation to limit my dreams or potential. A friend encouraged me to get behind the decks, and rather than politely decline, I said yes! It was time for me to take back my power and my life and take up space while spreading joy and sharing my passion for the music I'd adored for so long. I've been behind the decks ever since!

WHAT IS YOUR FAVORITE THING ABOUT YOUR JOB AND WHAT MADE YOU FALL IN LOVE WITH WHAT YOU DO?

Music has been one of my great loves, and I cherish my record collection. They've been my closest confidants over the years. To me, DJing and radio introduce friends to future friends. When I play a record, I share a story – love, heartbreak, pain, joy, liberation. For those 2 minutes and 46 seconds, we're able to forget about everything else, and we're free to focus on how we feel. The bass line is our heartbeat. The rhythm is our conscience. The groove is our lifeline. Music is a universal sound for us to heal, unite, and celebrate. At that moment, we're connected.

Elevating that experience through DJing is one of life's greatest pleasures. I love what I do!

WHAT ADVICE DO YOU HAVE FOR SOMEONE WANTING TO PURSUE THIS CAREER?

Be authentic. Be bold. Dig, dig, dig. Collect what speaks to you and your heart.

> " Be creative, find what makes you unique, and contribute that sparkle to the culture.

Dream big and be ambitious. The hustle for gigs and venues is hard but also exciting! It's an opportunity for you to sell yourself and your experience. Know your worth and believe in yourself. Surround yourself with people who inspire you to be better. Never stop listening. And go for it!

WHAT IS SOMETHING YOU WISH MORE PEOPLE KNEW ABOUT THE WORK YOU DO?

It takes time, effort, and a lot of heart to dig, research and curate DJ sets for them to have soul. DJing is a real expression and a reflection of the dance floor – a symbiotic relationship. Together we thrive.

However, the music industry is still deeply flawed. It's a male-dominated industry, and legitimately occupying space – not due to our looks or because we're a token choice – is a challenge.

As women in the industry, it's important we connect with other women of music and build safe environments to create, inspire and support one another. There's still so much progress to be made within the vinyl scenes – we have the responsibility and the honor of empowering ourselves and each other.

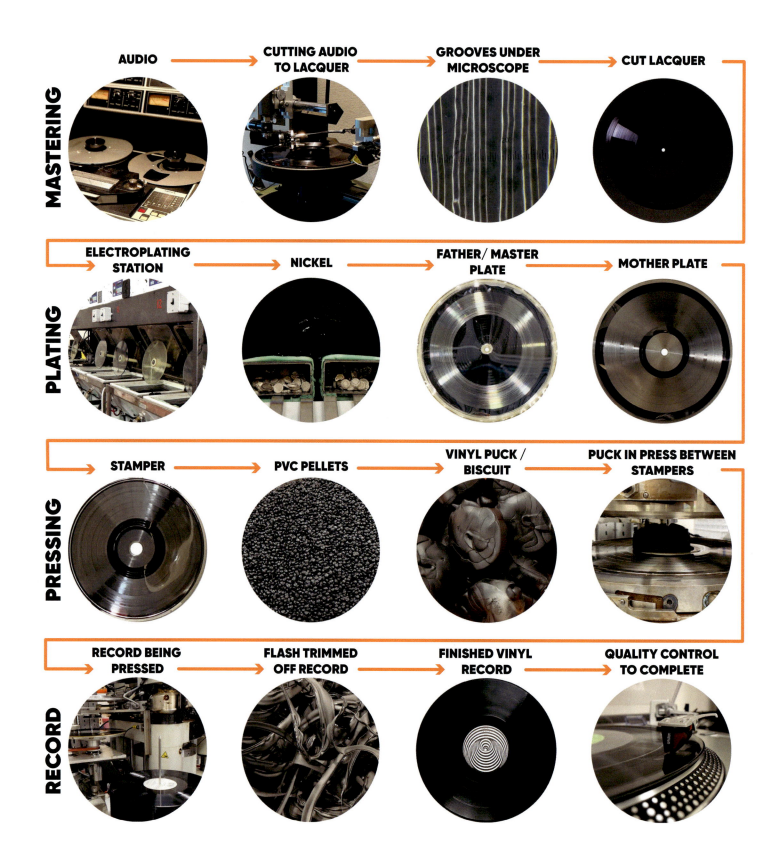

Glossary

33 1/3 rpm (Record Speed) – Rate of speed in which 33 and one-third rotations of a turntable platter are completed per 60-second duration. This is the speed best suited for playing microgroove LPs and is currently the standard for most 10" and 12" albums.

45 rpm (Record Speed) – Rate of speed in which 45 total rotations of a turntable platter are completed per 60-second duration. This speed is best suited for EPs pressed on 7" records, but some other specialty-pressed records utilize this speed as well.

78 rpm (Record Speed) – Rate of speed in which 78 total rotations of a turntable platter are completed per 60-second duration. This is the standard for most shellac resin recordings created up through the first half of the 1900s. These recordings hold approximately 5 minutes of sound per side.

Acetate – A lacquer that has not been electroplated to create stampers. These discs have a soft surface and are easily damaged. However, they do feature cut grooves, so they are playable on a turntable. See also: Dubplate, Reference Lacquer, Test Acetate, and Transcription Disc.

Bone Music – Russian bootleg lathe cut recordings after the second world war, cut into recycled x-ray film.

Carbon Black – Colorant additive used to create black vinyl that also adds durability and strength to the PVC mix.

Center Hole – The hole in the center of your record that sits on the spindle of your turntable. Formed by a center pin punching through the vinyl puck.

Center Label – The circular label appearing on each side of a pressed record displaying images, sides, rpm, song titles, and other recording information.

Colorants – Color additives used to create an endless rainbow of color and design options for record manufacturing.

Compact Discs (CDs) – Polycarbonate plastic discs with a reflective metallic digital data storage layer covered with an acrylic plastic coating that is read with a laser.

Deadwax – The unengraved area near the center label of a record. Referred to as "dead" because it contains no sound.

Debossed – Recessed ornamental pattern or image design on the surface of an object. This means the grooves at this stage are below the surface of the stamper, like on your records at home.

Digital Audio Workstation (DAW) – Electronic components and/or software used to record, edit, and produce audio files.

Direct Metal Mastering (DMM) – Process for transferring recorded sound that bypasses lacquer discs, relying on a copper-coated disc instead. A lathe engraves the audio directly into the copper disc. The copper disc goes right from the cutting process to plating, skipping the first silver spray in the electroplating process.

Dubplate – A lacquer that has not been electroplated to create stampers. These discs have a soft surface and are easily damaged. However, they do feature cut grooves, so they are playable on a turntable. See also: Acetate, Reference Lacquer, Test Acetate, and Transcription Disc.

Electroforming – A distinction from the generalized electroplating process, which coats one metal with another, electroforming creates a separate part or piece at the end of the process. See also: Electroplating.

Electroplating – A process that coats a metal object with another metal using an electrical current passed through a chemical solution. This process is used to create a stamper in the record manufacturing process.

Embossed – Raised ornamental pattern or image design on the surface of an object. This means the grooves at this stage are raised from the surface of the stamper rather than etched below the surface like on your records at home.

Extended Play (EP) – A record containing more tracks than a single but fewer than a full-length album. Introduced by RCA Victor as a 7" record playing at 45 rpm with about 7 minutes of (optimal) sound on each side.

Father Plate – The first nickel plate produced in the electroplating process which is a reverse copy of the original lacquer cut featuring debossed grooves, also referred to as a master plate. See also: One-Step (Electroplating Process).

Flash – The excess PVC that spills over the edges of a stamper on a record press during the pressing process. Flash is trimmed from the outer edge of a record before it is released from the press.

Galvanics – Generation and use of electrical current generated by a chemical reaction.

Gramophone – Synonym for Phonograph, Record Player, and Turntable.

Heat Stabilizers – An additive in the special PVC mix for record making. These additives help to make the overall compound more robust and neutralize hydrogen chloride gas.

Lacquer – An aluminum disc coated in a thick layer of nitrocellulose with a highly flammable mirror-like sheen. The first tangible element in the record manufacturing process, mastering engineers cut sound into a spiral-shaped groove flowing from the outer edge toward the center label.

Lathe-Cut Disc – A playable plastic record cut one at a time using a special rotating machine called a lathe. Used mostly for small quantity, collectible sound recordings.

Long Play (LP) – A record containing more tracks than an Extended Play (EP). Introduced by Columbia in 1948 and showcases the "microgroove" technology play at 33 1/3 rpm. LPs are 12" in diameter and provide about 21 minutes of (optimal) sound on each side.

Lubricants – An additive in the special PVC mix for record making. This additive assists the puck of PVC to spread more evenly and consistently across the surface of the stamper during production.

Master Plate – The first nickel plate produced in the electroplating process which is a reverse copy of the original lacquer cut featuring debossed grooves. See also: Father Plate, One-Step (Electroplating Process).

Microgroove Record – Innovation in record technology created by Peter Goldmark, allowing for longer durations of music to be cut into each side of the recorded disc. The grooves were read using a special Microgroove stylus (maximum tip radius 0.001 in or 25 µm) and provided 21 (optimal) playback time per side spinning at 33 1/3 rpms.

Mono – Sound recorded using one audio channel, intended to be heard as if it is coming from one position. Also known as Monaural or monophonic sound. Recordings made before 1958 are Monaural.

Mono Embossing – Modern lathe cut method of scratching the grooves into the plastic with the stylus. Final products tend to have lower fidelity, lower volume, and shallower grooves.

Mother Plate – A copy of the master or father plate made by repeating the electroplating process using the father plate as a base. The mother plate is a positive nickel casting that can be electroplated multiple times to create additional stampers usable on a record press. Each mother can produce approximately ten stampers. The grooves of a mother are embossed and are playable on a regular turntable. See also: Two-Step (Record Press Process).

Mould – The "holder" that stampers are placed onto when put inside the record press. They have channels inside them for steam and cold water to circulate through in order to heat and cool the PVC.

Noise Floor – The total sum of unwanted signals in audio; this is the lowest threshold of signal that is detectable in a recording. A lower noise floor means there is less undesirable or ambient noise in the playback of a recording, allowing a broader range of low sounds to be audible. If you hear someone say a recording is "dead quiet," this means it has a very low noise floor and sounds very clear.

Nonfill – A production error occurs when the PVC does not get hot enough in the record press, causing the material to inadequately fill the grooves on a stamper's surface.

One-Step (Electroplating Process) – The electroplating process that goes from a cut lacquer disc through one step of electroplating to create a single master stamper. That master stamper or father plate is used on the record press to stamp out discs. The term "one-step" refers to there being only one step from the original cut lacquer to the stamper. This technique is best for small runs of less than 500 copies. Also see: Father plate

Phonograph – Synonym for Gramophone, Record Player, and Turntable.

Plasticizers – Chemicals added to PVC to make the mix more pliable, resulting in a more flexible and more durable final product.

Polyvinyl Acetate (PVA) – One of the first additives mixed with PVC to create the special vinyl record blend.

Polyvinyl Chloride (PVC) – The foundational component of the vinyl record plastic mix. In its pure form, it is translucent white, and somewhat brittle.

Pre-echo – The phenomenon where an echo or "print-through" in a recording, mechanically induced by a manufacturing fault, is heard before the sound causing it when the recording is played. Caused by an overheated cutting stylus on the cutting lathe.

Pre-plate – Step in the electroplating process after silvering where the silvered lacquer is briefly placed into an electroplating bath of a lower temperature and lower amperage and coats the silver with an initial thin layer of nickel. This plating bath helps acclimate the lacquer to the electroplating process and eliminate pre-echo in the final stamper plates. See also: Pre-echo.

Proofing – The process of "proofreading" or looking over all final printed materials for errors in text or image assets.

Puck – The thick circle-shaped biscuit of vinyl mixture placed between two stampers on a record press is heated under immense pressure to flatten and fill all the grooves and create a record.

Record press – A machine used for manufacturing vinyl records by stamping PVC pucks or biscuits using a hydraulic press, heated by steam, and fit with thin nickel stampers and baked paper center labels. The compression of these parts fills the stamper grooves with hot PVC filling the grooves to create a finished record.

Rectifier – An electronic power supply device that converts available AC (alternating current) electricity to DC (direct current) electricity.

Reference Lacquer – A lacquer that has not been electroplated to create stampers. These discs have a soft surface and are easily damaged. However, they do feature cut grooves, so they are playable on a turntable. See also: Acetate, Dubplate, Test Acetate, and Transcription Disc.

Release Date – The anticipated date that a record is to be released.

Rotations Per Minute (rpm) – The measure of rotations. In terms of making records, it is the method of identifying the speed at which a turntable platter spins. Also known as Revolutions Per Minute.

Silvering – The first coating in the electroplating process of record stamper creation. A thin layer of silver is sprayed onto a deep-cleaned lacquer surface. This layer of silver acts as the base layer for growing the nickel plating during the electroplating bath. See also: Electroplating, Lacquer, Stamper.

Stamper – The plate used on a record press to form each side of a record. The hot vinyl puck is pressed between an A-side stamper and a B-side stamper to produce a two-sided recording.

Stereo – Sound recorded using two audio channels, intended to be heard as if it is coming from multiple directions. Also known as stereophonic sound. Records made during the 1960s can be either Stereo or Mono, but stereo eclipsed mono, becoming the dominant format in audio entertainment by 1968.

Stereo Diamond – Modern lathe-cut method of cutting the grooves into the plastic with a cutting stylus. Final products tend to have a higher fidelity cut with a deeper groove.

Stitching – A production error that occurs when the pressed PVC is cooled down too far, creating what looks like tiny lines or "stitches" on the surface of the record, creating an audible repetitive tearing sound.

Street Date – The date that a physical product will be available to consumers in stores.

Test Acetate – A lacquer that has not been electroplated to create stampers. These discs have a soft surface and are easily damaged. However, they do feature cut grooves, so they are playable on a turntable. See also: Acetate, Dubplate, Reference Lacquer, and Transcription Disc.

Test Pressings – The first records made from a set of stampers on a record press. These discs are reviewed by quality control specialists at the record pressing plant and sent to the artists and/or labels for final approval before the full order is fulfilled by the plant.

Three-Step (Electroplating Process) – The electroplating process best suited for large pressing runs of more than 10,000 copies. The three-step process involves setting aside both the first father plate and the first mother plate. These original plates are used to generate new additional copies (stampers) of its counterpart without requiring a new lacquer cut.

Transcription Disc – A lacquer that has not been electroplated to create stampers. These discs have a soft surface and are easily damaged. However, they do feature cut grooves, so they are playable on a turntable. See also: Acetate, Dubplate, Reference Lacquer, and Test Acetate.

Two-Step (Electroplating Process) – The electroplating process that relies on two steps of electroplating after the lacquer disc is cut by the mastering engineer. The first step of electroplating creates the father or master plate. A second round of electroplating creates a "mother" plate which can be used to make ten father stampers. The first father is used as a stamper during the two-step process. This process is best used on runs of less than 10,000 copies.

Vinyl Broker – This is a middle entity that works as a bridge between artists or labels and the pressing plant to handle the ordering and logistics of the manufacturing process.

Photo Credits

ABOUT THE AUTHOR

Jenn with some of her Black Sabbath Records and Cats – courtesy photo   xi

INTRODUCTION

© Women in Vinyl Logo   2

CHAPTER ONE: LACQUER CUTTING

Cutting Lathe at Sterling Sound by © Jenn D'Eugenio   6
Grooves Cut Into a Lacquer Under a Microscope by © Jenn D'Eugenio   8
Jett Galindo – by © Jei Romanes & Gaile Deoso – Wanderlust Creatives   10
Jett Galindo – by ©Jei Romanes & Gaile Deoso – Wanderlust Creatives   12
Jett Galindo – by © Jei Romanes & Gaile Deoso – Wanderlust Creatives   13
Amy Dragon – by © Emma Rose Browne   14
Amy Dragon – by © Emma Rose Browne   16
Amy Dragon – by © Emma Rose Browne   17
Heba Kadry – by © Jackie Roman   18
Heba Kadry – by © Jackie Roman   20
Heba Kadry – by © Jackie Roman   21
Margaret Luthar – courtesy photo   22
Margaret Luthar – courtesy photo   24
Mandy Parnell – by © Simon Weller   26
Mandy Parnell – by © Simon Weller   28

CHAPTER TWO: ELECTROPLATING

Stampers on a wall by © Jenn D'Eugenio   30
Image of the Electroplating Stations by © Jenn D'Eugenio   32
Plating Infographic by © Jenn D'Eugenio   35
Yoli Mara – by © Ashley Hylbert   36

Yoli Mara – by © Christopher Morley	38
Janine Lettmann – by © Lina Larsen	40
Janine Lettmann – by © Weronika Specht	42
Janine Lettmann – by © Weronika Specht	43
Desiree Oddi – by © Phara Jorgensen	44
Desiree Oddi – courtesy photo	46
Desiree Oddi – courtesy photo	47
Elsie Chadwick – by © Sharon Coke	48
Elsie Chadwick – by © Sharon Coke	50
Elsie Chadwick – by © Sharon Coke	51
Emily Skipper – courtesy photo	52
Emily Skipper – courtesy photo	54
Emily Skipper – courtesy photo	55

CHAPTER THREE: MANUFACTURING

Record Press at Memphis Record Pressing by © Casey Hilder	56
Caren Kelleher – by © Amanda Hoffman Art	62
Caren Kelleher – by © Amanda Hoffman Art	64
Caren Kelleher – by © Amanda Hoffman Art	65
Anouk Rijnders – by © Tim Knol	66
Anouk Rijnders – by © Sander van Dijk	68
Anouk Rijnders – by © Tim Knol	69
Ren Harcar – by © Bridget Caswell	70
Ren Harcar – courtesy photo	72
Ren Harcar – courtesy photo	73
Brianna Orozco – by © Catrina Traylor-Francis	74
Brianna Orozco – by © Catrina Traylor-Francis	76
Karen Emanuel – courtesy photo	78
Karen Emanuel – courtesy photo	80

CHAPTER FOUR: DISTRIBUTION

Distribution of old Distribution center at United Record Pressing by © Jenn D'Eugenio	82
Amanda Schutzman – courtesy photo	86
Amanda Schutzman – courtesy photo	88
Amanda Schutzman – courtesy photo	89
Jocelynn Pryor – courtesy photo	90
Jocelynn Pryor – courtesy photo	92
Christie Coyle – courtesy photo	94
Christie Coyle – courtesy photo	96

Shelly Westerhausen Worcel – courtesy photo 98
Shelly Westerhausen Worcel – courtesy photo 100
Shelly Westerhausen Worcel – courtesy photo 101

CHAPTER FIVE: RECORD LABELS

Lay Bare Recordings Releases by © Jenn D'Eugenio 102
Désirée Hanssen – by © Jack Tillmanns 106
Désirée Hanssen – by © Metal Skeleton Photography 108
Désirée Hanssen – by © Karin Correia Castilho edited by Jack Tillmanns 109
Katy Clove – courtesy photo 110
Katy Clove – courtesy photo 113
Riley Manion – courtesy photo 114
Riley Manion – courtesy photo 116
Riley Manion – courtesy photo 117
Katrina Frye – by © Karen Hernandez 118
Katrina Frye – by © Karen Hernandez 121
Julia Wilson – courtesy photo 122
Julia Wilson – by © Mclean Stephenson 124
Julia Wilson – courtesy photo 125

CHAPTER SIX: RECORD STORES

Erie Street Records by Owner © Sam Heaton 126
Lolo Reskin – by © Monica McGivern 130
Lolo Reskin – by © Monica McGivern 132
Lolo Reskin – courtesy photo 133
Brittany Benton – by © Amber N. Ford 134
Brittany Benton – courtesy photo 136
Brittany Benton – by © Manny Wallace 137
Sharon Seet – by © Jennifer SeetBeh 138
Sharon Seet – by © Jennifer SeetBeh 140
Sharon Seet – by © Jennifer SeetBeh 141
Claudia Wilson – by © George Howard 142
Claudia Wilson – courtesy photo 144
Claudia Wilson – by © George Howard 145
Shirani Rea – by © Noe Cugny 146
Shirani Rea – courtesy photo 148
Shirani Rea – by © Cedric Bloxson 149

CHAPTER SEVEN: LATHE CUTTING

A Vinyl Recorder T-560 by © Robyn Raymond	150
Robyn Raymond – by © Jordon Woolley	154
Robyn Raymond – by © Jordon Woolley	156
Robyn Raymond – by © Jordon Woolley	157
Bailey Moses – courtesy photo	158
Bailey Moses – courtesy photo	160
Bailey Moses – courtesy photo	161
Tasha Trigger – by © Adam Martin	162
Tasha Trigger – by © Adam Martin	164
Tasha Trigger – by © Adam Martin	165
Emily Nobumoto – by © Aaron Bobeck Photography	166
Emily Nobumoto – by © Aaron Bobeck Photography	168
Emily Nobumoto – courtesy photo	169
Oihane Follones – by © Bernd Jonkmannsm	170
Oihane Follones – by © Roberto Argüelles	172

CHAPTER EIGHT: DJING

Turntables by © Ray Blevins	174
Colleen Murphy – by © Ellie Koepke	178
Colleen Murphy – by © Adam Dewhurst	180
Misty Fujii – by © Ayana Wyse	182
Misty Fujii – by © Hipshakers, Mexico City	184
Misty Fujii – by © Nude Restaurant Kobe Japan	185
Monalisa Murray – by © Kris Perry	186
Monalisa Murray – by © Mousa Kraish	188
Monalisa Murray – by © Kai Sutton	189
Dana Brown – by © Erin Willis	190
Dana Brown – by © Erin Willis	192
DJ Honey – by © KF	194
DJ Honey – by © KF	196
DJ Honey – by © KF	197
Pressing Outline all images by © Jenn D'Eugenio	198

References and Further Reading

CHAPTER ONE: LACQUER CUTTING

Contributors to Wikimedia Projects. (2022, December 13). Acetate disc – Wikipedia. Retrieved from https://en.wikipedia.org/wiki/Acetate_disc
Contributors to Wikimedia Projects. (2022, June 25). Direct metal mastering – Wikipedia. Retrieved from https://en.wikipedia.org/wiki/Direct_metal_mastering
LACQUER MASTERING – United Record Pressing. (2018, February 20). Retrieved from www.urpressing.com/lacquer-mastering/
Merriam-Webster. (n.d.). Lacquer. In *Merriam-Webster.com dictionary*. Retrieved April 8, 2022, from www.merriam-webster.com/dictionary/lacquer
Vinyl 101: How to Make a Vinyl Record – Furnace Record Pressing. (2022, April 08). Retrieved from www.furnacemfg.com/how-to-make-vinyl-record/
Women in Vinyl Podcast - Vinyl 101 on Apple Podcasts. (2021, April 04). Retrieved from https://podcasts.apple.com/us/podcast/episode-two-tech-talk-vinyl-101/id1559469148?i=1000515718754

CHAPTER TWO: ELECTROPLATING

Contributors to Wikimedia Projects. (2022, June 25). Direct metal mastering – Wikipedia. Retrieved from https://en.wikipedia.org/wiki/Direct_metal_mastering
Merriam-Webster. (n.d.). Galvanic. In Merriam-Webster.com dictionary. Retrieved January 15, 2023, from www.merriam-webster.com/dictionary/galvanic
Vinyl 101: How to Make a Vinyl Record – Furnace Record Pressing. (2022, April 08). Retrieved from www.furnacemfg.com/how-to-make-vinyl-record/
Women in Vinyl Podcast – Vinyl 101 on Apple Podcasts. (2021, April 04). Retrieved from https://podcasts.apple.com/us/podcast/episode-two-tech-talk-vinyl-101/id1559469148?i=1000515718754

CHAPTER THREE: MANUFACTURING

Contributors to Wikimedia Projects. (2022, May 26). Record press – Wikipedia. Retrieved from https://en.wikipedia.org/wiki/Record_press
Hand Drawn Pressing – CompanyWeek. (2022, May 08). Retrieved from https://companyweek.com/article/hand-drawn-pressing#:~:text=But%20the%20comeback%20of%20vinyl,'50s%2C%22%20says%20Blocker
What Is Vinyl? This Is What Records Are Made Of | Vinyl Record Life. (2022, May 15). Retrieved from www.vinylrecordlife.com/this-is-what-vinyl-records-are-made-of/

CHAPTER FOUR: DISTRIBUTION

McDonald, H. (2020, August 13). What is music distribution? Liveaboutdotcom. Retrieved from www.liveabout.com/music-distribution-defined-2460499
The History of Music Distribution | Features | MN2S. (2020, September 04). Retrieved from https://mn2s.com/news/label-services/the-history-of-music-distribution/
Vinyl Lives: A History of Record Stores in America. (2020, November 22). Retrieved from www.vinylives.com/history.html

CHAPTER FIVE: RECORD LABELS

Contributors to Wikimedia Projects. (2022, December 18). Record label – Wikipedia. Retrieved from https://en.wikipedia.org/wiki/Record_label

Other Record Labels. (2022, June 08). Retrieved from www.otherrecordlabels.com/the-history-of-record-labels

CHAPTER SIX: RECORD STORES

Contributors to Wikimedia Projects. (2022, December 23). Record Store Day – Wikipedia. Retrieved from https://en.wikipedia.org/wiki/Record_Store_Day

Contributors to Wikimedia Projects. (2023, January 08). Record shop – Wikipedia. Retrieved from https://en.wikipedia.org/wiki/Record_shop#:~:text=are%20independent%20retailers.-,History, phonographs%2C%20cylinders%20and%20shellac%20discs.

Ediriwira, A. (2020). A comprehensive guide to grading vinyl records. Vinyl Factory. Retrieved from https://thevinylfactory.com/features/a-comprehensive-guide-to-grading-vinyl-records/

How To Grade Items. (2022, August 02). Discogs. Retrieved from https://support.discogs.com/hc/en-us/articles/360001566193-How-To-Grade-Items

Katz, R. (n.d.). A brief history of vinyl records. The Vinyl Revivers. Retrieved from https://thevinylrevivers.com/a-brief-history-of-vinyl-records/

Prince, P. (2020). Record grading 101: Understanding the Goldmine Grading Guide. Goldmine Magazine: Record Collector & Music Memorabilia. Retrieved from www.goldminemag.com/collector-resources/record-grading-101

Vinyl Lives: A History of Record Stores in America. (2020, November 22). Retrieved from www.vinylives.com/history.html

CHAPTER SEVEN: LATHE CUTTING

BONE MUSIC. (2022, August 20). Retrieved from www.x-rayaudio.com/x-rayaudiohistory

Lathe Cut Record vs Pressed Vinyl. (2022, August 20). Retrieved from www.tangibleformats.com/about.html#:~:text=Lathe%20cutting%20utilizes%20the%20same, sound%20into%20the%20blank%20disc.&text=This%20makes%20lathe%20cut%20records, for%20making%20affordable%20custom%20vinyl

LATHECUTS.com. (n.d.). What are lathe cuts? Retrieved from https://lathecuts.com/

Spice, A. (2016, May 04). A beginner's guide to lathe cutting your own records. The Vinyl Factory. Retrieved from https://thevinylfactory.com/features/a-beginners-guide-to-lathe-cutting-your-own-records/

The Secret Society of Lathe Trolls. (n.d.). Retrieved from http://lathetrolls.pbworks.com/w/page/22959575/The%20Secret%20Society%20of%20Lathe%20Trolls

CHAPTER EIGHT: DJING

Contributors to Wikimedia Projects. (2022, September 12). Francis Grasso – Wikipedia. Retrieved from https://en.wikipedia.org/wiki/Francis_Grasso

Contributors to Wikimedia Projects. (2022, November 26). Grandmaster Flash – Wikipedia. Retrieved from https://en.wikipedia.org/wiki/Grandmaster_Flash

McGroggan, S. (2019, February 11). Where did the term 'DJ' come from? WDIY. Retrieved from www.wdiy.org/arts/2019-02-11/where-did-the-term-dj-come-from

Michael, D. (2016). DJing: A (very) brief lesson in early history. Passionate DJ. Retrieved from https://passionatedj.com/djing-a-very-brief-lesson-in-early-history/

The Term "Disc Jockey", Since When? History of Djing. (2019, February 05). Skilz DJ Academy. Retrieved from www.skilzdjacademy.com/post/2019/02/01/dj-history#:~:text=History%20of%20DJing&text=In%201935%2C%20American%20radio%20commentator, Ancient%20Shepherds%20in%20Otley%2C%20England

Index

33 1/3 rpm 58, 198
45 rpm 58, 198
78 rpm 58, 198
A&R 104
A2IM 97; Black Independent Music Accelerator Program 120; mentorship programs 93
account representatives 95–96
acetate 9, 198
Alagirisamy, Alagiri 141
album artwork 61, 105
All Media Supply 87
Alliance Entertainment 92
American Decca Records 176
Amoeba Music 187–8
AMPED Distribution 91–92
Analog Vault, The (TAV) 139–140
apprenticeships 16
Artone Studio 68
audio engineer 23–25
Avatar Studios 12

Bakery, The 12
Barnes, Jerry 12
Basic Disc Mastering 165
Beat Junkie Institute Of Sound 189
Beat Junkies Crew 189
Benton, Brittany 134–137
Bertus Distribution 68
biscuit *see* puck
Block, Martin 176
BMG 104
Boden, Larry 165
Bone Music 153, 198
bootlegs 152–153
Borders Books & Music 92
Brain Drain 123
Brittany's Record Shop 135–136
Brown, Dana 190–193
Brown, Violet 92
Brunswick Records 176

Canada Boy Vinyl 156
Capitol Records 104
carbon black 59, 198
carbon footprint 59

Cascade Record Pressing 16
cassettes 29, 60
CBS 104
center hole 34, 198
center label 9, 59, 104, 152, 198
Chadwick, Elsie 48–51
chemicals 58–59; cyanide 32; hydrogen chloride 58; nickel sulfamate 32; nitrocellulose 8; polyvinyl chloride 58–60, 200
Chicago Mastering Service 23–24
Classic Album Sundays 179–181
Clove, Katy 110–113
colorants 59, 198; carbon black 59, 198
Columbia Records 58, 84, 104
compact discs (CDs) 29, 43, 67, 128, 173, 198
conferences 39, 68
Coyle, Christie 94–97
cyanide 32

Dark Sky Mastering 23
Dead Oceans 115
deadwax 9, 198
debossed 33, 198
Digital Audio Workstation (DAW) 21, 198
Direct Metal Mastering (DMM) 9, 34, 198
Discogs 128–129
distribution 84–85; marketing 91–93, 99–101, 141, 156; sales representative 60–61, 87–89
Dixon, Mike 157, 160–161
DJ Dusk 188
DJ Francis Grasso 176
DJ Honey 194–197
DJ Kool Herc 176
DJ Misty 182–185
DJ Red-I 134–137
DJs 5, 80, 116, 128, 132, 136–137, 144, 171, 176–177; *see also individuals*
Doornroosje 108
Dragon, Amy 14–17
drink sum wtr 115
dubplate 9, 198
Dudley, Mary Dee 173

eco-friendliness 59
Edison, Thomas 128

Edison Records 84, 104, 128
electroforming 32, 198
electroplating 9, 32–34, 37–38, 41–42, 61, 198; one-step process 33–35, 199–200; plating tech 45–47; three-step process 33–35, 201; two-step process 33, 35, 201
electroplating QC 41–43
Emanuel, Karen 78–80
embossed 33, 198–199; mono embossing 153, 199
EMI 104
Erler, Helmut 173
extended play (EP) 58, 199

father plate 33–35, 41, 61, 199
Fitzgerald, Kevin 188
flash 60–61, 199
Follones, Oihane 170–173
For the Record 140
Frye, Katrina 118–121
Fujii, Misty 182–185
Furnace Record Pressing 4

Galindo, Jett 10–13
Galloway Studio 107
galvanics 50–51, 53–55, 199
galvanics technician 53–55
gender bias 4
Ghostly International 115
Giordano, Cory 156
Girls Rock 116
gold records 63
Gold Rush Vinyl 63–65
Goldmark, Peter 58
Goldmine 129
Gonsalves, Adam 16
Gotta Groove Records 71–72
Grammy U mentorship program 93
gramophone 104, 128, 199
Grandmaster Flash 176

Haagsma, Robert 108
Hamilton 60
Hanssen, Désirée 106–109
Harcar, Ren 70–73
heat stabilizers 58, 199

HITS ACT 93
Hocus Bogus Records 159
Hopkowitz, Gladys 173
Human Machine Interface (HMI) 60
hydrogen chloride 58

Inner Ocean Records 156
internships 12, 21, 100, 116, 168
iZotope 21

Jagjaguwar 115

Kadry, Heba 18–21
Kaplan, Mitch 133
Kapp, Jack 176
Kaufmann, Flo 173
Kelleher, Caren 62–65
Key Production Group 79–80
Keyes, Karrie 13
KPL All-Stars 187

lacquer 4, 8, 199
lacquer cutting 8–9, 32, 38, 61, 152, 172; *see also* mastering engineer
Ladies Of Sound 187
lathe cutting 8–9, 38, 152–153, 155–157, 159–161, 163–165, 167–169, 171–173, 199
Lathe to the Grave 163
Lathe Trolls 165
Lathesville 171–172
Lauretta Records 119
Lay Bare Recordings 107
Lee, Vicious 188
Leesta Vall Sound Recordings 167–168
Lettmann, Janine 40–43
Linna, Miriam 173
long play (LP) 58, 128, 199
lubricants 58, 199
Luthar, Margaret 22–25
Luv, Mark 188

machines *see* record press
Manalo, Aji 13
Mancuso, David 180
Mango Landin' Bar 144
Manion, Riley 114–117
Manor Residential Studio 28
manufacturing 58–61, 65, 153
Mara, Yoli 36–39
marketing 91–93, 99–101, 141, 156
master plate 199; *see also* father plate
mastering engineer 11–13, 15–17, 19–21, 23–25, 27–29
Mastering Lab, The 12
Matsushita 177
Memphis Record Pressing (MRP) 75
mentors 4, 12, 16, 25, 29, 64, 93, 108, 169
Merge Records 111–112
microgroove record 58, 199
Midnight Choir 123
Mischief Managed 119

mono 58, 152–153, 199
mono embossing 153, 199
Moses, Bailey 158–60
mother plate 33–35, 61, 199
Motown 104
mould 59, 199
Moving the Needle 79, 81
Murphy, Colleen "Cosmo" 178–181
Murray, Monalisa 186–189
Mutterstecherin 41–43

Napster 84
Needle + Groove Records 87
Nice Rights 123
nickel sulfamate 32
Nickol, Brian 77
Night Beat Records 183
nitrocellulose 8
Nobumoto, Emily 166–169
noise floor 59, 199
nonfill 61, 199
NPR 23–25
Numero Group 115–116

Obata, Shuichi 177
Oddi, Desiree 44–47
office manager 49–51
one-step electroplating process 33–35, 199–200
Orozco, Brianna 74–77
Ow-Young, Eugene 140

Pallas 41–43
Parnell, Mandy 13, 26–29
Passion for Vinyl 68, 108
Peaches Records 147
Pheenix Alpha 60
Phillips 128
phonograph 128, 200
plasticizers 58, 200
plating tech 45–47
polyethylene terephthalate glycol 152
Polygram 104
polyvinyl acetate (PVA) 58, 200
polyvinyl chloride (PVC) 58–61, 200
pre-echo 32, 200
pre-plate 32, 200
Press On Vinyl 53
Pressing & Distribution (P&D) Agreements 85
Presto 6N 160
Pride, Alan 107
Prism DJs 187
production director 115–117
production manager 111–113
production supervisor 75–77
proofing 61, 200
Proper, Darcy 13
ProTools 152
Pryor, Jocelynn 90–93
puck 59–61, 200
Pure Vinyl Record Shop 143

qualifications 12, 24, 28, 100, 112, 168
quality control/assurance 41–43, 45, 61, 71–73
quantity 60
Queer Vinyl Collective (QVC) 191–192

RAP ACT 93
Raymond, Robyn 154–157
RCA Victor 58, 104, 128, 199
Rea, Shirani 146–149
Reaper 21
Record Industry 67–69
record labels 104–105; *see also individual labels*
record press 59–60, 200; press innovation 60
Record Store Day 128
record stores 128–129; *see also individual stores*
Record Technology Inc. (RTI) 45–46
Recording Industry Association of America (RIAA) 8, 84
Recording Workshop 20
rectifier 32, 200
Red Spade Records 155
Redeye 95–97
reference lacquer 9, 200
release date 105, 200
Reskin, Lolo 130–133
Rice Is Nice 123
Rijnders, Anouk 66–69
Robinson, Sylvia 173
Rock, Pete 188
rotations per minute (rpm) 58, 200; 33 1/3 rpm 58, 198; 45 rpm 58, 198; 78 rpm 58, 198
Rough Trade Distribution 80
royalties 85, 105

Saddest Factory Records 115
sales representative 60–61, 87–89
Samplitude 21
Savile, Jimmy 176
Sax, Doug 12
School of Audio Engineering 28
Schutzman, Amanda 86–89
Secretly Canadian 115
Secretly Group 115–16
Seet, Sharon 138–141
Sefchick, Ian 23
senior mastering engineer 27–29
shellac 58, 84, 128
silvering 32, 200
Skipper, Emily 52–55
Slipped Disc Records 88
Slovenly Recordings 171–172
social media 4, 60, 97, 104, 128
Sonicscoop 21
Sony Music 68, 104
Sound of Niche (SoN) 107
SoundGirls 12
Spiller, Henry 128
Spillers Records 128
Spinderella 188
Staff, Ray 29
stamper 32–35, 49–51, 53, 58–61, 200; father plate 33–35, 41, 61, 199; mother plate 33–35, 61, 199

Stamper Discs 49–50
Stein, Linda 173
stereo 58, 153, 200
stereo diamond 153, 200
stitching 61, 200
street date 105, 200
Sub Pop 104
Sukate, Martin 173
Sun Records 104
Super D 92
Sweat Records 131–132

Tape Op audio conference 20
TAV Records 139–141
test acetate 9, 200
test pressings 61, 200
three-step electroplating process 33–35, 201
Tinnemans, Manuel 108
Toolex Alpha 60
Toronto Soul Club 183–185
transcription disc 9, 201
Transworld 92
Trident Studios 29
Trigger, Tasha 162–165
two-step electroplating process 33, 35, 201

Umoja Hi-Fi Soundsystem 187
Universal Music Group 104

Vermeulen, Mieke 69
Vermeulen, Ton 68–69
vinyl broker 105, 201
Vinyl Record Manufacturing Association 39
Vinyl Revolution Record Shows 87–88
Virgin Megastore 132
Viryl Technologies 60

"War of the Speeds" 58
WarmTone 60
Warner Brothers 104
Wavelab 21
Wax Mage Records 71–72
weight 60
Welcome to 1979 23–24, 37
Westerhausen, Shelly 98–101
Wherehouse Music 92
Wilson, Claudia 142–145
Wilson, Julia 122–125
Winchell, Walter 176
Women in Vinyl 4–5, 13, 43, 51, 61, 97, 155, 173
Women's Audio Mission 21
Worrall, Dan 21
Wowsville 171–172

YouTube 21, 64